Implementation Project Management

How to Manage B2B Software and Hardware Product Implementations

CORNELL TRIBBET

Copyright © 2020 Cornell Tribbet

All rights reserved. No part of this publication may be reproduced, distributed, or transmitted in any form or by any means, including photocopying, recording, or other electronic or mechanical methods, without the prior written permission of the publisher, except in the case of brief quotations embodied in reviews and certain other non-commercial uses permitted by copyright law.

Although the author and publisher have made every effort to ensure that the information in this book was correct at press time, the author and publisher do not assume and hereby disclaim any liability to any party for any loss, damage, or disruption caused by errors or omissions, whether such errors or omissions result from negligence, accident, or any other cause.

Cover Design by 100Covers.com
Interior Design by FormattedBooks.com

ISBN: 978-1-7350733-1-6 (Paperback)
ISBN: 978-1-7350733-0-9 (Ebook)

TABLE OF CONTENTS

INTRODUCTION	V
SECTION 1: THE BACKSTORY	**1**
CHAPTER 1: WHAT A WEEK!	3
ANKLE BITER	3
THE PEOPLE AT THE BOTTOM	5
CHAPTER 2: CLASS REUNION	9
ZO WAS "SHARP" AND "SMART"	11
THE ULTIMATUM	13
CHAPTER 3: A NEW PERSPECTIVE	15
THE UNIVERSE, IN ITS INFINITE WISDOM, PROVIDED A GUIDE	16
MY EYES WERE OPENED	18
SECTION 2: PROJECT MANAGEMENT FOUNDATION	**23**
CHAPTER 4: KEY DEFINITIONS	25
WHAT IS A PRODUCT?	25
WHAT IS A PROJECT?	26
WHAT IS PROJECT MANAGEMENT?	26
WHAT IS A SUCCESSFUL PROJECT?	26
WHAT IS A FAILED PROJECT?	26

CHAPTER 5: KEY ROLES AND RESPONSIBILITIES	29
SALES	29
SPONSORS	30
STAKEHOLDERS	30
SUBJECT MATTER EXPERTS	30
TECHNICAL LEADS	31
PROJECT MANAGERS	31
SECTION 3: THE CIPM FRAMEWORK	**33**
CHAPTER 6: OVERVIEW OF THE CIPM FRAMEWORK	35
CHAPTER 7: INITIATE	43
CHAPTER 8: PLAN	53
CHAPTER 9: EXECUTE–MONITOR–CONTROL	71
CHAPTER 10: CLOSE	89
SECTION 4: CONCLUSION	**97**
SECTION 5: EXTRAS	**101**
FREQUENTLY ASKED QUESTIONS	101
PRODUCTIVITY MYTH BUSTERS	103
GLOSSARY	107
ACKNOWLEDGEMENTS	113

INTRODUCTION

70% FEWER ACTIVITIES ARE NEEDED FOR PRODUCT IMPLEMENTATION PROJECTS COMPARED TO TRADITIONAL PROJECTS

> **PRODUCT**: Commercial, pre-packaged software and/or hardware that does not require significant configuration or customization for each business customer. Typically implemented by a professional services or customer success team.

You've likely arrived at this book because you want to learn about managing product implementation projects. Perhaps you are brand new to managing implementations and are looking for a resource to help you get started. Or perhaps you lead product implementation projects already and you're looking for a new or better way to manage them.

Depending on how long you've been searching, you've likely discovered, or soon will discover, there isn't much information available on managing product implementations.

The majority of project management books are geared towards large, generic projects. As a result, in those books you'll find many activities, terms, and concepts that are not applicable to product implementations.

Even experienced project managers can find these traditional project management approaches challenging! So, it makes sense that the complexity of most project management frameworks causes many product companies to delay offering implementation/professional services to their customers for as long as possible.

Fortunately, I've discovered a better way to manage product implementations — a way that addresses the unique needs of implementation projects. Instead of spending your time wading through terms, methodologies and concepts that just don't apply, we'll focus on an easy-to-follow, practical framework that you can start using immediately to bring your implementation projects to a successful close.

This practical how-to guide will help:

- Individuals with little to no implementation project management experience learn how to deliver successful implementations
- Experienced implementation project managers who want to be more productive
- Business-to-Business (B2B) product companies that want to jumpstart their enterprise implementation services

In this book, readers will discover a simple, scalable framework for customer-facing product implementations, along with:

- Uncommon advice and life lessons
- Things to think about
- Productivity myth busters
- Funny, real-life short stories
- "How would you handle it?" scenarios

Who is this book for?

This book contains many golden nuggets of information that will benefit anyone involved in project management.

For the following at a software or hardware product company, this book is an information goldmine:

- Beginner and experienced implementation project managers
- Consulting and Customer Success team members involved in implementations
- Managers of implementation project managers
- Senior managers considering an implementation services offering

I've poured what I've learned from managing 500+ successful implementation projects over the last 20 years into this book, and hope you'll find yourself doing one or more of the below while reading it:

- If you're new to implementation project management, feeling an ever-increasing level of confidence that you can successfully manage implementation projects even with limited/no prior experience
- If you're an experienced implementation project manager, reconsidering the way you think about and manage implementations
- Nodding in agreement
- Laughing out loud
- Smiling

Ready to get started? You have two options:

1. Start with **SECTION 1**: THE BACKSTORY for a **humorous look at my journey** to develop an optimized project management approach. It wasn't a smooth ride. In fact, just like a lot of projects, there were unrealistic expectations, miscommunications, and team conflicts along the way!

2. Start with **SECTION 2**: PROJECT MANAGEMENT FOUNDATION if you want to jump right into the **project management topics**.

SECTION 1
THE BACKSTORY

I'd been living in the Chicago South Loop and had recently started a new job with the same title as my last few—project manager. Life was good. The one thing different was that I'd shifted from managing information technology (IT) security projects to software product implementations.

Before taking the job, I was sure:

1. I had excellent project management skills.
2. My career was the priority for the next few years.
3. At least 80% of the processes in any type of project would be similar, whether in IT, security, construction, digital, manufacturing, research, infrastructure, event, legal, etc.
4. As an excellent project manager, it would only take me a few months to master the 20% of processes that are unique to product implementations.

You'll see in Section 1 that I was wrong about all four!

The stories in Section 1 (and throughout the book) are based on true events. The names, locations, and some details were changed to protect individual identities. The made-up names are based on characters from science fiction TV shows and movies. Watch for the name origin notes (with *) at the end of the story/chapter where I introduce these characters.

- Chapter 1: What a Week! – Two events made me start to question my project management skills.

- Chapter 2: Class Reunion – I met the woman who made me want to work less.

- Chapter 3: A New Perspective – A mentor helped me to understand why my current project management approach was not working, and what I needed to do instead.

CHAPTER 1
WHAT A WEEK!

Awareness of what we do well and what we need to improve on at work is a major challenge for most.

How does a project manager go from total confidence to confused and humbled in a matter of days? Well, for me, it began with two project status meetings.

ANKLE BITER

It's a Tuesday afternoon, and I'm in a weekly client project status meeting in Chicago's Near North Side.

The agenda? Reviewing progress from the last meeting, issues, and upcoming activities. Our meetings are typically limited to a handful of key Project team members. For this meeting, several new team members have joined as we're starting a new project phase. Their presence seems to bring a renewed sense of energy to the large conference room.

Sara,* the project sponsor, starts by introducing the new team. Next, she introduces the original team members.

Looking at me, she announces, "Cornell is our ankle biter."

All of the original team members start laughing. Sara half-smiles, but not a laugh. It's an "I practiced this joke in front of the mirror a hundred times, I'm so happy it worked" kind of smile.

I don't smile or laugh because I have no clue what an ankle biter is. Judging by the puzzled looks on the new team members' faces, they're also clueless.

After about five to ten seconds, the laughter starts to die down.

With a matter-of-fact tone, I say, "OK, let's get started on today's agenda."

The laughter starts up again, and I'm even more confused. I'm sure the laughing has something to do with the ankle biter comment, but I'm not sure as I still don't know what an ankle biter is.

After the second wave of laughter stops, we work our way through the agenda items in short order and end the meeting.

As Sara and I are walking out of the conference room, I ask about the ankle biter reference.

> "Hey, Sara. What's an ankle biter?"
>
> "You know, those little dogs that nip at your ankles until you give 'em what they want."
>
> "Wow. Is that how you see me?"
>
> "Cornell, you're an OK project manager. If I didn't like you, I wouldn't joke around with you. See you next week."

IMPLEMENTATION PROJECT MANAGEMENT

Sara's words caught me off-guard. I didn't understand why she thought I was only doing an OK job. The project was ahead of schedule, on-budget, and the team had only worked on items listed in the contract.

* Name Origin: The project sponsor reminded me of Sarah Connor (portrayed by Linda Hamilton) in the *Terminator* movies. Sarah did not laugh in *Terminator: Dark Fate*, and I don't recall seeing or hearing the project sponsor laugh in all my three months working with her.

THE PEOPLE AT THE BOTTOM

Fast-forward to Thursday of the same week, and I'm in Milwaukee for a weekly project status meeting with a different customer. It's a fairly large implementation. We normally require 90 minutes to get through the recently completed activities, issues, and upcoming activities. The project is slightly behind schedule, so I press team members to commit to revised completion dates for their open tasks.

The meeting ends, and as I walk through the conference room door, I notice the database administrator (DBA) assigned to the project walking towards me.

"Hey Steven.** What's up?"

"Cornell, I did not appreciate you trying to embarrass me."

"Um, what'd I do wrong?"

"You asked me when my action items would be done in three different ways. I'm very busy, and don't have much time for this project. You're just a project manager, you don't understand the importance of what DBAs do."

"*Just* a project manager?"

"Not to be offensive, but you're pretty much at the bottom of the IT hierarchy."

"I didn't know there was an IT hierarchy."

"The people at the bottom never do. DBAs are at the top because we're responsible for the company's crown jewels [*data*], followed by admins [*network, server, security, etc.*], and then helpdesk and project managers."

I half-jokingly ask, "Where do IT managers fall in the hierarchy?"

He responds with a straight face, his voice monotone, "They are on the same level as DBAs."

I smile. "Thanks for helping me understand I'm at the bottom of the hierarchy."

I'm still feeling the sting of "*just* a project manager" and now want to end this fictional IT hierarchy game.

IMPLEMENTATION PROJECT MANAGEMENT

"One question, Steven."

"What?"

"Will you have your late action items completed before next Thursday's status meeting?"

He turns and walks away, shaking his head.

Later, while walking to my car, I started replaying my verbal battle with Steven, and laughed out loud. I wasn't laughing because I'd won the battle. Quite the opposite. Steven wiped the floor with me. It was because there were no witnesses to the beat down. He chose to confront me in private instead of during the meeting. I was happy there wouldn't be any Steven versus Cornell jokes or stories next week.

I spent most of my Friday afternoon drive from Milwaukee back to Chicago thinking about the post-meeting discussions with Sara and Steven. I'd been with this product company for three months, and previously managed IT and security projects for five years. This was really the first time I'd given any thought about whether I was a good project manager.

7

Once back in Chicago, after a few minutes of going through emails, I remembered my high school reunion was on Saturday. It was also Labor Day weekend.

** Name Origin: Steven reminded me of Steve Rodgers (aka Captain America, portrayed by Chris Evans) from the Marvel *Avengers* movies. He is one of, if not the best, hand-to-hand fighters on the planet, and Steven hit me with some great verbal shots.

THINGS TO THINK ABOUT

Self-awareness can be a challenge

Awareness of what we do well and what we need to improve on at work is a major challenge for most. Managers tend to overrate their staff at appraisal time. As individuals, we tend to overestimate the quality of our work. In my head, I was doing an excellent job as an implementation project manager. However, the conversations with Sara and Steven suggested I was not. It's a good idea to occasionally check with your coworkers to see if there is something you should be doing differently.

There is an IT hierarchy

My 2020 internet searches—"IT hierarchy" and "computer jobs hierarchy"—confirmed Steven was correct about there being a hierarchy, but wrong about the rankings. Project managers and IT managers are considered mid-level management. Database administrators are considered lower-level, which I disagree with. From my perspective, the hierarchy is irrelevant. It typically takes a team effort to deliver a successful project. There's no time for one group to think they are more important than another.

CHAPTER 2
CLASS REUNION

It's amazing how two words can take someone from no life to having a life.

The post-meeting discussions with Sara and Steven got me thinking about whether I needed to adjust my project management approach, but it was Zo* who made me take action.

Here's how I met Zo.

It's Saturday morning, and I have just finished packing for a couple of days, including a really sharp gray suit bought for a special occasion. My high school reunion is three hours away in Springfield, Illinois.

Fast-forward to the Saturday night dinner dance.

I arrive at the class reunion fashionably late. Outside the ballroom, there is a group of classmates who I haven't seen in four or five years. After talking with them for about ten or 15 minutes, I hear someone in the distance ask,

"Is that Cornell in the gray suit?" I turn around and see a beautiful woman walking towards me with a big smile. My first thought is, "She looks like an actress or a model." My second: "This new suit is already paying dividends."

She's a few steps away. I smile, take a step towards her, and say, "Hi."

She catches me off-guard with a closer, more intimate, and longer hug than I expected. While still hugging me close, she whispers, "Hi Cornell..." *I had no idea those two words would be the start of something special.*

As we separate, she says, "I haven't seen you in at least 10 years. How've you been?"

I still have no clue who she is. So I glance down for a split second to read her name tag.

"What? Are you looking at my name tag? Do you not know who I am?"

Beautiful and perceptive, I now think. With the smoothness of actor Denzel Washington, I reply,

> "Of course I know who you are, Zo; just wanted to check your last name to see if you were still married."

> "No, I'm not married."

My bad week just got better. Still channeling Denzel, I delivered a sincere-sounding, but fake line.

> "Ohhhh, really sorry to hear that."

> "Are you still married?"

> "Ah, no, no, me too. I'm divorced too."

OK, I admit. That was not done with Denzel-like smoothness.

Zo said, "Wow ... I can't believe we grew up a block apart, went to the same grade school and high school, but we hardly ever talked."

"Yeah, I don't know why that was. Let's grab a table. I'd really like to hear what you've been up to."

OK, I took the lead by suggesting we grab a table and offered to listen instead of talk about myself. I'm back in Denzel-mode.

We talk pretty much the rest of the dinner-dance. As we're exchanging phone numbers at the end of the night, she asks, "Are you coming to any of tomorrow's events?"

"Unfortunately not. I have to work."

"But it's Labor Day weekend."

"I'm flying to Minneapolis on Monday and need to create a PowerPoint for the meeting and catch up on my other projects."

The look on Zo's face suggested she didn't understand why I was working on the weekend or traveling for work on a holiday. Also, I don't think she was used to a guy not wanting to spend time with her.

ZO WAS "SHARP" AND "SMART"

I called Zo a few days later, and it was another great conversation. It would have been good to see her in person right after reconnecting at the reunion, but at the time my focus was work and she lived three hours away. So, we talked on the phone. I'd never been one for long phone conversations, but I talked to Zo for hours almost every night. We talked about everything and anything: grade school, middle school, high school, careers, bucket lists and dreams, family and friends, likes and dislikes.

Even though the conversations were great and there was a strong connection, we didn't actually go out until a month after the reunion.

Zo came to visit me in Chicago. It was an incredible weekend! It seemed like the time from the Saturday morning pickup at Chicago Union Station to her Sunday evening train departure went by in a matter of hours. We packed a lot of stuff into a day and a half: Field Museum, coffee shops, jazz club, cool restaurants, a walk in Grant Park. Much to my surprise, I hardly thought about work.

She was sharp, smart, and had a good sense of humor.

We continued seeing each other every other weekend. My workload prevented us from being together more often. Even on the weekends we were together, I still worked at least a few hours on both Saturday and Sunday.

* Name Origin: I used the name Zo because she reminds me of Zoë Alleyne Washburne (portrayed by Gina Torres) in the 2005 movie *Serenity*. Both were smart, confident, kind, caring, and gorgeous.

THE ULTIMATUM

Zo wanted to spend more time with me. She was tired of hearing about the need to send out meeting notes within 24 hours, preparing for project kickoff meetings, updating project plans, escalated issues, and helping the Sales team win deals.

She gave me an ultimatum after a couple of months of every-other-weekend visits, "You need to find a way for us to spend more time together or—"

She stopped short of finishing the sentence, but I got the message. In truth, I hadn't really taken much action in reducing my hours up to this point. I enjoyed my job as it was, but I definitely did not want to lose Zo. So, I took her ultimatum seriously.

Reducing travel was the easiest way for me to recover time. I liked meeting face-to-face with customers and consultants, but it wasn't required. I would typically spend two to four days onsite with customers and consultants most weeks, spending the night out of town at least two days a week. A shift to conference calls allowed my monthly travel to go from ten to 12 days to three or four days, and I was still able to keep my projects on track.

Reducing my days onsite was a good move, as it allowed me to spend every weekend and holidays with Zo. My stress went down and my projects stayed on track, but I was still working 50 to 60 hours a week.

THINGS TO THINK ABOUT

Don't let work get in the way of having a life.

Looking at how and why I was working so much also made me realize what I was giving up. Going to Springfield every other weekend to see Zo also meant I saw my parents, siblings, nieces, nephews, and longtime friends more in a couple of months than in a normal year. I also started working out again. Life is short. Few people on their deathbed wish they had worked more; quite the opposite. One of the top deathbed regrets is having worked so hard that it negatively affected relationships.

Understand the difference between sharp and smart.

I described Zo as "sharp" and "smart." How well do you understand the difference between the two?

"Sharp" is something a person is born with. For example, they notice little things that others don't, they easily and quickly adapt to situations they haven't experienced before, and they seem to do everything well.

"Smart," on the other hand, isn't only related to your level of intelligence. It also has a lot to do with things that are learned. Additional learning reinforces natural intelligence, and that's why someone comes across as smart. So, you hear of people who are "book smart," "school smart," or "street smart."

Rare is the individual who is both sharp and smart. If you come across them at work or outside of work, try to stay connected with them in some way.

CHAPTER 3
A NEW PERSPECTIVE

It's not the things you don't know that most impact your life. It's the things you know to be true that are not true.

After receiving Zo's ultimatum, I spent weeks trying to figure out how I could work fewer hours and still successfully manage my projects. Reducing my travel time helped, but I was still working 50 to 60 hours a week. I spent most of my time on administrative tasks like scheduling and running meetings, writing up detailed meeting notes, and updating project documentation. Also, I spent a good amount of time working with Sales on activities such as responding to requests for proposals, estimating consulting hours required for proposals and contracts, and taking part in pre-sales calls.

I read lots of books and visited countless websites to learn more about personal productivity and time management. There was lots of generic information, but very little was applicable to implementation project managers.

Other project managers at my company were of little help. I shadowed three of my company's project managers when I joined the company. I was doing what they were doing. There were a few useful discoveries from my project manager talks. One, I was working more hours than other project managers. Two, I had more projects than anyone else. Three, no other project managers were assisting the Sales team with pre-sales activities.

I spoke with my manager about my number of projects and time spent assisting Sales. He agreed to not assign me new projects for *a while*. He requested I continue assisting Sales while he explored other options. Not receiving a new project for a couple of weeks would help me for the near-term but would not significantly reduce the number of hours I would work in the long-term.

Working fewer than 50 hours per week at my current job did not seem doable. I reached out to my recruiter and asked her to keep an eye out for remote IT project manager jobs with an 8-5 schedule.

THE UNIVERSE, IN ITS INFINITE WISDOM, PROVIDED A GUIDE

I had failed on my goal to only work 40 to 50 hours a week. Or so I thought . . . until I met the project manager who showed me the way.

Morph* was his name.

He had worked at the company for three years. My manager asked me to manage Morph's projects during his upcoming two-week vacation. Funny thing is, a week earlier I'd talked to my manager about reducing my workload, and now he wanted me to take on Morph's projects.

I hadn't talked with Morph before because other project managers had warned me about him during my first week on the job. They considered him a loser because his approach to managing projects differed from theirs. Morph primarily managed each of his projects with a Microsoft Excel spreadsheet and email updates. The other project managers took pride in creating and maintaining detailed implementation project plans, detailed meeting notes, and sending out detailed status reports.

I scheduled an hour session with Morph so that I could come up to speed on his projects before his vacation started.

"Hi Morph. Let's get started since we only have an hour."

He laughed, "You could have scheduled the call for 10 minutes."

Before I could respond, he reminded me that we hadn't met or spoken before.

We spent several minutes talking about how things were going at work before I shifted back into project manager mode. I asked him to send project docs to help bring me up to speed. He laughed again.

"I'll send you an email summary for my projects after our call."

"I'm concerned that there aren't project docs to help bring me up to speed and track project status while you're out."

He laughed once again. "You won't need them." *No wonder the other project managers consider him a loser.*

"I don't know how I can manage your projects without project plans, meeting note—"

He cut me off, "I have a status call in 15 minutes. Are you available? It'll help you understand why you won't need docs to manage my projects while I'm out."

Morph sent me the meeting invite, and I joined the call.

What happened next amazed me.

Morph and the Customer Project team members spent several minutes talking and laughing about stuff that had nothing to do with the project. He then spent a few minutes providing a verbal project status update and ended the call.

Morph called me right after. I answered, and as I was raising the phone to my ear, I could hear him laughing.

> "Are you still concerned about managing my projects without all of *your* project management docs?"

I was more confused than concerned. I didn't understand why the customer was good to end the call without going over the project plan or action items. Rather than try to figure it out, I just asked Morph if we could do lunch the next day. He laughed.

> "My calendar is pretty flexible. We can meet at the office."

*Name Origin: I chose the name Morph because he reminded me of Morpheus from the *Matrix* movies. Both were incredible mentors who saw the world differently than most.

MY EYES WERE OPENED

The next day, I got to the office around 6.30am to beat the highway traffic and to catch up on my work so that a long lunch with Morph was doable.

I'm not sure what time Morph got to the office, but I first saw him around 11.30am talking to our manager. I finished up my call, grabbed my notebook and pen, and tracked down Morph. We started the walk to my favorite Thai restaurant at around 11.45am. I spent the next two hours enjoying a good meal, listening to Morph's project management philosophy, and answering and asking lots of questions.

He started off asking questions about how I saw my job as a project manager; why I believed customers expected lots of project documentation; how much time I spent on documentation versus helping customers and consultants. It seemed like each answer resulted in him laughing. Looking back, I don't think he was laughing at my answers as much as my beliefs on managing implementation projects.

Then he went into detail about how he manages implementation projects and why he felt his approach was more effective than the heavy-documentation approach used by most other project managers.

If I had to summarize our conversation into a single sentence: he helped me understand why traditional project management processes were overkill for business-to-business product implementations. Morph was consistently delivering successful implementations but spending a lot less time on documentation than I did. Could this be the key to me having more time to spend with Zo?

On my drive home, I couldn't stop thinking about my lunch conversation with Morph. Is it really possible to manage a successful implementation without spending hours and hours on project plans, meeting notes, and status reports? Had I been doing everything wrong?

Morph's perspective made sense to me, but I couldn't just abandon my project management training and beliefs after a couple of conversations. However, it only took a few weeks for me to realize he was 100% right.

I gradually replaced detailed PDF meeting notes with email summaries and replaced detailed Microsoft Project plans with high-level Excel Project plans. Much to my surprise, not one customer asked about the changes. With this implied customer confirmation of Morph's "less project documentation is better" approach, I started focusing my efforts on becoming more of a help to customers and consultants, and less of an ankle biter—and spent more time with Zo.

I've spent the last 20 years building on what Morph taught me, and I've never regretted the decision to embrace a different approach to managing my implementation projects. This customer- and communication-focused approach allowed me to balance work and personal life.

In the next section, I'll define key terms and project roles that will help you better understand the customer-facing implementation project management (CIPM) framework.

THINGS TO THINK ABOUT

Form your own opinions.

I could have saved myself a lot of time to spend with Zo if I'd talked to Morph when I first joined the company. Instead, I listened to the negativity from other project managers. Talk with your peers and Project team members to form your own opinions of them rather than relying solely on the opinions of others.

It's not the things you don't know that most impact your life.

It's the things you know to be true that are not true. Before starting my new job, I believed all of the following to be true. Within a span of a few months, they were all proved wrong. We must continually seek feedback from others to grow.

False Truth #1: I had excellent project management skills.

Post-meeting conversations with Sara and Steven, lunch with Morph, and no complaints from customers after eliminating the detailed project docs were enlightening. I was not an excellent implementation project manager.

False Truth #2: My career was the priority for the next few years.

Strengthening my relationship with Zo was now my priority.

False Truth #3: At least 80% of the processes in any type of project would be similar, whether in IT, security, construction, digital, manufacturing, research, infrastructure, event, legal, etc.

Morph helped me understand implementation projects are different than most other types of projects. In Chapter 6, we'll start discussing why they are different.

False Truth #4: As an excellent project manager, it would only take me a few months to master the 20% of processes that are unique to product implementations.

It took a couple of years before I really felt 100% comfortable that I had a good approach for delivering successful implementations. After 20 years, I'm still tweaking my approach.

SECTION 2

PROJECT MANAGEMENT FOUNDATION

As you saw in Section 1, I didn't create a project management framework overnight. It took experimentation and adjustment over time, as I completed projects and learned what worked well and what didn't work so well. But it doesn't have to be an uphill climb for you. In the remainder of the book, we'll take a close look at each phase in the Customer-facing Implementation Project Management (CIPM) framework, as well as the accompanying activities.

Before we dive into the details of the CIPM Framework, we'll build a solid foundation, reviewing the key terms and key players common to most implementation projects. Each reader has different project management learning and work experiences. This section acts as a knowledge leveler to help ensure there is a shared understanding of key definitions and roles.

Starting in Chapter 4, we'll review key definitions for some of the basic terms that will help you better absorb the details as we move through the CIPM Framework.

Then, in Chapter 5, we'll take a look at key roles and responsibilities: who are the key players, what are they responsible for, and how they work together in implementation projects.

CHAPTER 4
KEY DEFINITIONS

Implementation project management is different! Whether you're new to project management or you're new to implementation project management, it's important to understand what the following key words and phrases mean.

This chapter covers a handful of definitions. The definitions here may differ slightly from those found in other books or on websites because of our implementation focus. You can also find these definitions, and those from other chapters, in the Glossary section at the end of this book.

WHAT IS A PRODUCT?

Commercial, pre-packaged software and/or hardware that does not require significant configuration or customization for each business customer. Typically implemented by a Professional Services or Customer Success team.

WHAT IS A PROJECT?

Any contracted planning-deployment-implementation-assessment-architecture review-optimization-knowledge transfer work assigned to the professional services/customer success group. (i.e., not considered a part of onboarding, training, or support.)

WHAT IS PROJECT MANAGEMENT?

The planning and managing of activities related to completing a project.

WHAT IS A SUCCESSFUL PROJECT?

Project success is difficult to define because it can differ from company to company. Success could be illustrated by one or more of the below:

- Project completed by the planned date, within budget, and only the planned activities were worked on
- How closely the project followed your company's project management process
- Customer management and end-users' happiness level
- High customer satisfaction survey scores
- Customer is willing to act as a reference to other potential customers

WHAT IS A FAILED PROJECT?

Factors such as not finishing on time, working more consulting hours than planned, working on items not listed in the contract, or not meeting customer expectations are not desired but don't necessarily mean a project failed. Other factors such as whether the customer is using the product or has purchased additional licenses/different products can outweigh issues encountered during the project.

A failure is when the project is canceled due to an issue the Implementation team could not overcome or when the customer badmouths a company and its products to other businesses. This usually happens due to one of the following:

- Incorrect expectations set during the sales process
- Bugs (i.e., the product is not doing what's expected or not responding in the expected way)
- Customer leadership changes, who prefer a different product
- Unsupported customer environments

Based on my experience and discussions with many other project managers, I estimate fewer than 2% of product implementations are failures.

THINGS TO THINK ABOUT

These concise definitions are meant to provide a foundation for the coming chapters.

I recommend web search engines (e.g., Google, Bing) and *A Guide to the Project Management Body of Knowledge, PMBOK® Guide*, Sixth Edition | Newton Square, PA | Project Management Institute, 2017) to fill in definition gaps after reading this book.

Memorization is not required.

Don't feel like you need to memorize these definitions or spend lots of time researching. They'll become clearer as you read through the book.

CHAPTER 5
KEY ROLES AND RESPONSIBILITIES

Product implementations are typically made up of a core team of fewer than 10 individuals. This small size allows for fast decisions and the flexibility to schedule meetings with relatively short notice. Additional team members from both the customer and your company are pulled onto the project as needed.

To better understand how this team works together, let's look at the core Project team members and their responsibilities in the course of a typical implementation project.

SALES

Sales team members are responsible for pre-qualifying customers, pre-sales education, identifying requirements, providing price quotes, and closing deals. The titles vary by company but include account director, account executive, account manager, sales engineer, systems engineer, and solutions architect.

After the deal closes, the Sales team typically stays involved to help ensure a smooth transition, identify additional opportunities, and continue managing the company-customer relationship.

SPONSORS

Internal (your company) and customer project sponsors are typically members of senior or executive management. They are the project's final decision-makers, help escalate and resolve issues, free up staff to work on projects when they are assigned elsewhere and have ultimate responsibility for the project. Also, the customer project sponsor typically funds the purchase of product and implementation services.

STAKEHOLDERS

An individual, group, or organization that may affect, be affected by, or perceive itself to be affected by a decision, activity, or outcome of a project. (PMBOK®, Sixth Edition.)

SUBJECT MATTER EXPERTS

Individuals from your company and customers with technical expertise who provide guidance in a specific area and perform project tasks related to their area of expertise (e.g., architect, backup administrator, cloud system engineer, customer support specialist, database administrator, network engineer, security engineer.)

Note: Throughout the book we refer to subject matter experts who work at your company as consultants.

TECHNICAL LEADS

Typically, your company and the customer will assign technical leads to jointly provide technical expertise, develop and confirm architecture, interface with Project team members, create technical diagrams, and ensure the environment is ready for installation/configuration.

Note: Depending on product complexity, the consultant may act as both technical lead and consultant.

PROJECT MANAGERS

You, the implementation project manager, and the customer-assigned project manager work together to plan, schedule, and manage day-to-day project activities. The customer project manager typically has a limited role in product implementations, as the implementation project manager has overall accountability for the project's success and activities.

THINGS TO THINK ABOUT

Do you want to be an implementation project manager?

If you're not already an implementation project manager or thinking about it, consider implementation project management as a career.

Over my 30-year IT career, I've worked for many different companies and in many different roles. I've done Unix administration, project management, management, and sales. Implementation project management is the one that I've focused on for the last 20 years. The decision was easy: management without the human resources component, every day is different from the one before, good pay, surrounded by people smarter than me, smile and laugh multiple times every workday, and an in-demand job skill.

Do you need a project management professional certification?

While I am a certified Project Management Professional (PMP), I've only been asked twice in the last 10 years whether I was certified, both times by recently certified project managers. If you're not considering project management as a career, then certification may not be worth pursuing. If you're looking at project management as a career, then I would recommend certification, as hiring managers often list certification as a requirement.

SECTION 3

THE CIPM FRAMEWORK

Most project management books tend to be hundreds of pages long because they cover lots of topics in great depth. We'll take a different approach with the Customer-facing Implementation Project Management (CIPM) Framework. Because this is a streamlined approach to project management, we'll focus on the project management activities that are primarily responsible for successful implementations.

That means we won't cover technical skills like how to run a great meeting, create concise emails, put together an attention-holding PowerPoint deck, or deep dive into project management theories. There are excellent books and websites that already cover these types of topics. We also won't cover product-specific items, since these are unique to each company.

Instead, we'll look at the specific, organized implementation steps that take a customer-facing product implementation project from initiation to closeout, using the CIPM Framework.

Use the framework to develop a consistent delivery process, which helps facilitate best practices and continual optimization. You can also adapt the framework based on your company's products, target customers, and your personal project management style. I've provided concise yet thorough explanations for each element of the frame, so you can better choose which items to adapt or adjust to meet your unique needs.

Additionally, each of the chapters contains key terms and activities, challenges, things to think about, and a "how would you handle it?" scenario so you can flex your own project management skills.

In Chapter 6: Overview, we'll take a high-level look at how the CIPM Framework works, so you can understand the overall process from start to finish.

In Chapter 7: Initiate, you'll learn how to start an implementation project off on the right foot so that you establish a solid foundation with your internal team before engaging with the customer.

In Chapter 8: Plan, I'll lay out exactly what information to gather so you can create an effective project plan.

Then, in Chapter 9: Execute-Monitor-Control, we'll cover what the day-to-day management of a typical implementation project looks like, where to focus your time, and how to manage issues that might arise in the course of the project.

Finally, in Chapter 10: Close, you'll learn how to close out a project, ensuring your customer is well taken care of and any loose ends are tied up.

By the time you've completed this section, you should have a thorough understanding of the CIPM Framework. You'll also be prepared with solutions to some of the common challenges you might face. Most importantly, you'll be fully equipped to begin applying the CIPM Framework to real-life implementation projects.

CHAPTER 6
OVERVIEW OF THE CIPM FRAMEWORK

Do we really need another project management framework After all, we have PMBOK, Waterfall, Agile, Scrum, and PRINCE2, to name a handful!

The answer? Yes.

Those project management frameworks might work well with traditional and software projects, but implementation project management is different.

To successfully run implementation projects, we don't need as much documentation. Remember Morph from my story earlier in the book? Unlike the other project managers at my company, he was managing projects with only a spreadsheet, yet he was a successful implementation project manager.

I started off with a documentation-focused approach, and I was quickly overwhelmed, spending hours upon hours updating documents. Eventually, I saw the light and started tackling project management from a new angle. The CIPM Framework is the result—a streamlined, optimized way of managing implementation projects.

In this chapter, we dig into the CIPM Framework, examining in-depth what it is and why we need it.

KEY TERMS

Agile: A framework that is primarily used for software development projects. Agile projects typically consist of small, regular releases of software that are based on end-users' feedback, rather than larger and less frequent releases.

Best Practices: What successful companies of similar size and industry are doing.

Roadblock: Something that is stopping or slowing completion of a task.

Scope/in scope: Specific tasks the Implementation team will perform as noted in the Statement of Work

Simultaneous projects: A mix of projects in various stages (e.g., started, not yet started, paused.)

Standard/traditional project: The mythical project you typically read about in project management books. A relatively small percentage of project managers rarely, if ever, see or manage one. The scope is huge, has a seven-figure budget, 30 dedicated team members, and is the number-one priority for everyone in the company. I'm exaggerating a lot to make a point, but the majority of project management advice assumes you want info on running large projects at a large company.

Statement of Work (SOW): Your company provides the buyer with a document that details the project activities, timeline, tasks, hours, expenses, acceptance criteria, payment schedule, etc. It's a binding contract to be signed by your company and the customer.

Throw someone under the bus: To blame another person or group for something that went wrong.

How would you handle it?

Working with "difficult" people

As you read this chapter, think about how you would respond to the situation below. At the end of the chapter, you'll find out how I handled it and why.

Your manager has just assigned you a new project and let you know that Hano,* the customer project manager, was a real pain on two earlier projects. He suggests you don't take any nonsense from him and make it clear who is running the project on your first call. How would you handle it?

Why do we need an implementation-specific framework?

There are several differences between traditional projects and implementation projects:

Traditional Projects	Implementation Projects
Project details such as scope definition, estimated hours, acceptance criteria, and costs are finalized after the project is approved.	The SOW addresses many of the traditional project management activities including scope definition, estimated hours, acceptance criteria, and costs.
Typically, longer, more complex, and requires ten or more team members.	Typically, shorter, less complex, and require fewer team members.
The key project success factor is finishing on time, in scope and on budget.	The key project success factor is ensuring the customer sees business value in the product and has a positive implementation experience.
Project managers typically manage eight or fewer projects simultaneously.	Implementation project managers typically manage 10 or more projects simultaneously.

What is the CIPM Framework?

The CIPM Framework is designed to address the challenges unique to implementation project management. It lays out the steps that take a business-to-business (B2B), customer-facing implementation project from initiation to closeout and can be adapted based on the situation or need.

The framework helps ensure Project teams are delivering projects in a consistent manner. This, in turn, helps facilitate best practices and continual optimization.

Throughout the remaining chapters in this section, we'll follow this CIPM framework and steps, which are similar to the PMBOK® (Project Management Body of Knowledge, Sixth Edition) process groups.

- Initiate (Chapter 7)
- Plan (Chapter 8)
- Execute-Monitor-Control (Chapter 9)
- Close (Chapter 10)

KEY ACTIVITIES

Chapters 7 through 10 cover the step-by-step activities of each phase in detail. In addition to these, there are three keys to managing successful projects with the CIPM Framework. The rest of the framework doesn't work without these three keys, so make sure to keep them at the forefront of every implementation project you manage!

Know the status

As the project manager, being aware of how your projects are progressing, current challenges, and upcoming activities is a must. Management and team members will expect you to know this information. Plan to get regular updates via email, meetings, and phone conversations to stay informed and on top of the status at all times.

Communicate, Communicate, Communicate

As a project manager, you should spend the majority of your time on communication-related activities such as meetings, verbal and email updates, and phone conversations. Communicating in a proactive, clear, and concise manner keeps everyone on the team in sync and is critical to the success of the project.

Keep things moving

It's inevitable that an unexpected issue slows or brings a project to a pause. If the project is slowed or paused for an extended period, team members tend to get pulled onto other projects and activities. If a project you're managing slows down or comes to a halt, take action to keep things moving! Identify activities that were planned for later in the project that can be started earlier to maintain momentum. Then prioritize eliminating the cause of the roadblock.

TOP 3 CHALLENGES

1. **Your company's management expects a traditional project management approach**

    Many project management experts and books promote a documentation-heavy approach, but this is based on mythical projects that don't reflect the reality of most product implementation projects.

How I handle it

For most projects, I provide enough documentation to satisfy management. For high-profile customers, I increase the level of documentation, as there are more eyes on the project.

2. Difficulty finding implementation-specific information

I'm always on the lookout for books and websites related to product implementations. Unfortunately, there are very few resources available.

How I handle it

I talk to other implementation project managers to share ideas, find out what they're doing, what tools they're using, and what they're working on.

3. Juggling many projects

As mentioned earlier in this chapter, implementation projects are typically shorter, less complex, and require fewer team members. So, the expectation is that implementation project managers will simultaneously manage a double-digit number of projects.

How I handle it

When I get above 25 to 30 projects, it's a challenge for me to stay on top of them all. I let my manager know when my workload is high, and request no additional projects until a few projects close out.

THINGS TO THINK ABOUT

The CIPM Framework does not work for all project types.

It's appropriate for software and hardware products that do not require significant configuration or customization. Other types of projects should leverage a different project management framework.

CIPM versus Agile

Implementation projects typically have a fixed scope and budget, and the schedule is somewhat flexible.

Agile projects use an approach that continually refines scope, with the schedule and budget fixed. Creating a SOW without a predefined scope would be a major challenge for product companies. However, you can easily include some Agile components in your implementations:

Stand up: A meeting focused on Project team members answering three questions similar to:

1. What have you completed since the last meeting?
2. What will you work on until the next meeting?
3. What are your roadblocks, if any?

Sprints: Time-boxed periods, typically two weeks, when a team works to complete a set of predetermined tasks.

Sprint planning: Identifying the tasks to be completed in a sprint.

How would you handle it?

Working with "difficult" people

Remember the scenario from the beginning of the chapter? Your manager assigned you a new project with a customer project manager who was a pain on two earlier projects.

Here's how I handled it

Something I've internalized since I met Morph was not to pre-judge someone. On my first call with the customer project manager, Hano, I mentioned an update from one of my team members that there were challenges in the last project, and that I'd like to understand what my company could do better on this project. He spent 30 minutes giving me his perspective on the difficulties of working with my company, and what he would like to see done differently on this project.

Most of Hano's requests were reasonable, and we discussed those that were not doable. The project completed without issue, and he sent an email to my manager and the Account team letting them know the implementation "met his expectation." *What? You were expecting a glowing email?*

* Name Origin: Hano expected perfection much like Thanos of the Marvel *Avengers* movies. Like Thanos, Hano made it clear what he expected, and you never wanted to disappoint as his response would be disproportionate.

CHAPTER 7
INITIATE

The surest way to ensure a project completes successfully is to start it successfully.

This chapter focuses on understanding what was sold, why the customer purchased it, and gathering information that helps ensure a productive project kickoff meeting.

We'll cover the key steps, including reviewing the Statement of Work and Purchase Order, completing the internal preparation call, calling the customer project sponsor, and creating a project folder.

KEY TERMS

Action items: Small tasks assigned to project team members. Typically identified during meetings and/or associated with completion of a project plan task.

Critical success criteria/factors: The key activities that must go well for the project to be considered a success from the customer's perspective.

Expectations: Things a customer expects but are not documented in the SOW.

Fixed price (Fixed): The contracted payment amount does not change regardless of the hours worked by your company to complete the project or to reach certain points within the project.

Kickoff: The project team's first formal meeting.

Prerequisites (pre-reqs): Activities to be completed or information provided prior to product installation or configuration.

Purchase Order (PO): The buyer sends your company a request to order a product or service. The document includes but is not limited to products, services, descriptions, quantities, prices, and payment terms. The PO and SOW contain similar information; however, the SOW is more detailed.

Quote: The quote provides customers with information needed to generate a PO (e.g., quantities, descriptions, address, contact.)

Sanitize: Removing proprietary/confidential/sensitive information and data from a document.

Time and Materials (T&M): Customer is billed based on the Project team members' contracted hours or days spent working on the project.

How would you handle it?

Custom documentation

As you read this chapter, think about how you would respond to the following situation. At the end of the chapter, you'll find out what happened, what I learned, and my recommendations for handling custom documentation requests.

On my initial call with Vade, the project sponsor, he mentioned the uniqueness of their environment and requested custom documentation. I recommended we discuss the specifics on the kickoff call.

During the kickoff call, several team members confirmed the need for custom documentation because of their unique environment. I asked for the specifics; however, the customer was unable to provide them. They just knew standard docs would not work for their environment.

This is a common request. Customers see install manuals, administrator guides, and troubleshooting documents with hundreds of pages that are not applicable to their environment. So, they ask the Implementation team to create documents specific to their environment.

How would you respond to this "custom documentation" request during the meeting?

KEY ACTIVITIES

Review the SOW/PO/Quote

Obtain a high-level understanding of the project and identify disconnects. A couple of examples: scope looks like it will take 100+ consulting hours; however, the customer purchased 40 hours. There is a reference to onsite work, but there's no reference to expenses. This review prepares the project manager for the internal preparation call with Sales.

Schedule and complete the Internal Preparation Call

This is a call with your company's Sales team to discuss the primary reason the customer purchased, what was purchased, how soon the customer wants to get started, critical success factors, other sales opportunities within the account, and anything else the Implementation team should know about the deal or customer. This call prepares the project manager for the customer sponsor call.

Note: If a partner company was involved in the sale, include their Sales team members in this call.

After the call, a Sales team member should send an email to the customer introducing the project manager and setting the expectation that the project manager will be contacting the sponsor shortly to discuss next steps.

Call project sponsor

The project manager calls the project sponsor. The easiest way for a project manager to make a great first impression is to have a good understanding of what was purchased and why it was purchased. This is the reason why you should complete the internal prep call before calling the sponsor.

This call typically lasts 10 to 20 minutes. Call topics should minimally include critical success factors, readiness for the kickoff meeting, kickoff meeting objectives, who the project manager should work with to schedule the kickoff, and the anticipated project start timeframe.

Note: The project sponsor may direct you to another team member for this call (e.g., the CIO may be the project sponsor, but prefer not to be directly involved in the implementation.)

Create a project folder

Create an email/online/shared folder for each customer to make organizing and searching for specific information easier. I recommend creating the folder with the customer name.

TOP 3 CHALLENGES

1. Customers or Account teams often want to start the project before the SOW is signed.

How I handle it

I typically push back on starting projects before the SOW is signed. However, there are times when my Management team wants to start a project early due to customer or Account team pressure.

When faced with this situation, I'll begin working with the customer on prerequisites. I set the expectation that additional pre-reqs may be identified after the consultant engages.

2. Not being able to bill all "Initiate Phase" T&M hours

You may spend a one to three hours on the "Initiate" phase. Some customers will be uncomfortable approving consulting hours prior to speaking with you.

How I handle it

During the "Customer Project Sponsor" call, I summarize to-date activities. This summary and keeping billable hours under two hours tends to eliminate billing pushback.

3. Undocumented environment changes between the time sales discussions started and when the SOW is signed.

How I handle it

This is normal. Project managers should not assume customer-environment info is current. The pre-sales to contracting process can sometimes take months, and customer environments are continually changing.

I look to identify environment changes during the initial project sponsor call and the kickoff meeting discussed in Chapter 8.

THINGS TO THINK ABOUT

Some customers require a project charter/definition/statement document before beginning a project.

This is the customer's internal, simplified version of a SOW. On occasion, a customer will ask the implementation project manager to create it. Offer to assist, but do not take ownership of this task. It sets a bad precedent when you take ownership of tasks outside of the Statement of Work at the beginning of a project.

PROJECT CHARTER			
Project Name	Andromeda Colony 12 – Weapons Platform		
Project Description	Deploy a defensive weapons platform in geosynchronous orbit		
Expected Start Date	8/10/2224	**Expected End Date**	9/18/2224
Project Sponsor	Colony Commander Gilat	**Risk Manager**	Colony Lead Accountant Filas
Business Case		**Expected Benefits**	
• Space pirates continue to test Colony defenses • Space worms have eaten two shuttles over the last 17 months		• Act as deterrent and defense against pirate ships • Discourage worms from feeding on shuttles • Increase colony confidence	
APPROVAL SIGNATURE			
Project Sponsor		**Date**	

Before accepting a new project, project managers should ensure they have the time to manage another one.

Project managers may have to work with their manager to transition one or more existing projects to a different project manager to ensure there is sufficient time for the new project.

Don't ignore warning signs.

If the project sponsor is difficult to reach or the project manager seems to regularly get upset on calls, those are warnings that should be addressed in a timely manner. Make your manager and the Account team aware of the warning signs and discuss with them how best to address the concerns.

Your primary customer contact might not have the official title or role of project manager.

This is not a problem. Regardless of the primary contact's title or role, what's key is making sure they have the time to perform anticipated customer project management duties. (e.g., managing completion of pre-reqs) If not, then let the Account team and your manager know there is a potential issue.

How would you handle it?

Custom Documentation

Remember Vade, the project sponsor who requested custom documentation on the kickoff call? The customer wasn't exactly sure what they needed, but they knew standard documentation wasn't going to work for their environment. Here's what happened next.

Here's how I handled it

There's little value in continuing to ask a question that a customer cannot yet answer on a kickoff call. I recommended that I follow up with Prin,** the customer project manager, after the call to obtain specifics and examples of what was being requested.

So, after the call, I sent an email to Prin suggesting she follow up with her Technical team for specifics on what was needed. I also included my company's standard installation and administration guides in the email.

Prin called the next day to let me know there was no need for custom documentation based on a discussion with her technical lead.

I sent an email to Prin summarizing our discussion and confirming that her team would not require custom documentation.

Vade joined our weekly status call for the first time a few weeks after the kickoff and asked how the custom documentation was coming along.

I waited a few seconds for Prin to answer the question ...

Silence.

I said, "Yes, I recall the customer documentation discussion."

I paused again for a few seconds to give Prin an opportunity to update Vade on her technical team's confirmation that custom documentation was not required. More silence.

It was me who messed up here, not Prin. I should have copied Vade on the email to Prin, confirming her team's decision that custom documentation was not required.

I started into what seemed like a 10-minute apology, but was probably only 15-20 seconds.

Vade cut me off with, "I appreciate the apology, but how are you going to make sure the custom documentation that I want gets created?"

I let him know our consultant would start working with the Project team to create custom documentation drafts, send the drafts to his team for feedback the following week, and going forward an update on custom documentation would be included in the weekly status report.

What that exchange taught me was that customers appreciate an apology, but they care more about how to prevent whatever happened from happening again. Since that call, when there is a need to apologize for me or my team, I make it quick and move on to prevention going forward.

Prin called me after the status meeting and thanked me for not throwing her under the bus. She confided in me that Vade is very unforgiving and difficult to work for.

The account manager, who I'd worked with on many successful projects, called me the next day. Vade called her after the status meeting and asked if I was the right manager for their project. The account manager let the sponsor know I was the right person. The project had no other issues from that point through closeout.

Because I messed up by not copying the project sponsor on my email to Prin and was now very familiar with the customer environment, I immediately agreed to provide custom documentation.

Below is the process I usually follow when asked for custom documentation:

1. Let Project team members know most customers find our standard documentation sufficient, and that our consultant will work directly with their team members during the implementation.
2. If a customer still wants custom documentation, I ask if there is specific documentation they are looking for. Based on the answer, I send their team members the relevant docs or links to docs and

ask them to let me know if our standard documentation is not sufficient.
3. If a customer still wants custom documentation, I request they email me sanitized examples of what other vendors have provided, or more details on what they are requesting.

   a) If it's something that can be completed in less than a couple of hours, I work with a consultant to create the documentation.

   b) If more than a couple of hours will be needed, the type of project determines the next step. If T&M, I let the customer know what can likely be provided based on available project hours. If Fixed and more than a couple of hours, I push back on the custom documentation but offer to assist by identifying existing company documentation that can be leveraged for their documents and reviewing documentation created by their team.

* Name Origin: The deep voice and the way Vade's team feared him reminded me of masked *Star Wars* character Darth Vader, voiced by James Earl Jones.

** Name Origin: The project manager reminded me of Princess Leia Organa in *Star Wars: New Hope*. Darth Vader tortured her during an extended interrogation, but she provided no useful information.

CHAPTER 8
PLAN

In Chapter 7, we learned how to obtain a high-level understanding of the project. In this chapter, we're going to take our data gathering to the next level, confirming the how, who, and when of the project.

You'll use the information you gather in this step to create a high-level project plan, outlining the key tasks and estimated completion dates. Capturing these items in a streamlined project plan helps ensure the entire team is on the same page without creating unnecessarily detailed documentation that is time-consuming to create and update.

KEY TERMS

Appliance: A self-contained device, sold as a product with integrated hardware and software.

Deliverables: Tangible things that will be delivered to the customer during the project. For example, project plan, test plan, working configuration, training, reports.

Issue: A problem that requires assistance from someone outside of the core Project team.

Knowledge transfer: Less formal than training, delivered by a consultant versus a trained instructor, and specific to each customers' environment.

Onsite: Project team members are physically working at a customer's office, data center, or a meeting location.

Remote: Project team members can work from anywhere (e.g., your company's office, consultant's home, coffee shop)

How would you handle it?

Out-of-scope requests

Occasionally, customers will request things that are outside the scope of the project's SOW. This can happen at any stage of the project, and it's important to know how to manage these requests.

As you read this chapter, think about how you would respond to the situation below. At the end of the chapter, I'll share with you how this situation played out in real life and a simple set of questions to consider when addressing out-of-scope customer requests.

Omer, a project manager at a Fortune 50 company asks you to manage the pre-req activities. He's looking for you to schedule meetings with his Technical team, track the tasks, and provide weekly status reports. The SOW clearly states pre-reqs are the customer's responsibility. How would you handle the project manager's request?

KEY ACTIVITIES

Schedule kickoff meeting

Work with the customer project manager or primary contact to schedule a kickoff meeting date and time. I typically schedule a one-hour meeting.

Prepare for kickoff

Work with the customer project manager to prepare pre-kickoff and kickoff meeting documents. These might include pre-reqs, data gathering questionnaires about their environment, request for names and titles of kickoff meeting participants, and the proposed kickoff agenda.

Send info docs in advance

Before the kickoff, send the customer project manager pre-meeting documents. Sending these prior to the kickoff makes for a more productive meeting.

Complete kickoff meeting

The kickoff meeting's primary purpose is to help ensure Project team members have a shared understanding of the project. Use a slide deck (e.g. PowerPoint, Google Slides) to help keep participants on topic, but try not to shut down productive non-agenda topic discussions unless they start to get too technical or lengthy.

Below are the minimum topics to cover:

> **Roles and responsibilities**: Team members job title/role, name, and contact info.
> **Scope**: What T&M and/or fixed services were purchased.

- If T&M, cover the number of hours and expenses for the project.
- If Fixed, there's no need to explicitly mention estimated hours or days, as this is indirectly addressed by the project plan.

Onsite and/or remote activities: Where will implementation activities be performed?

The product: Discuss or demonstrate how the product works at a high-level, as some kickoff participants may not be familiar with the product.

Change management process: What's the process for approving changes to the production environment?

Timeline/schedule: Resist the urge to commit to specific dates at the start of a project, as you might not yet have complete and accurate details. Use phrases like: "target dates," "anticipated timeline," and, "proposed schedule."

Testing: Which individuals/groups will do testing; what will be tested (e.g., is there any expectation of after hours, weekend, disaster recovery, live environment or performance testing?)

Readiness: Does network or server hardware have to be obtained; availability of team members assigned to the project.

If the above topics are covered before the scheduled time is reached, spend additional time reviewing:

Pre-kickoff questionnaire: open questions or clarifications, if needed.
Prerequisites: questions and estimated completion date.
Environment details: deep dive into existing architecture, hardware, applications, etc.

Send kickoff meeting notes

Email the meeting participants and stakeholders a summary of topics discussed, individual action items, and the slide deck. This helps to reinforce kickoff discussions, decisions, and action items ownership.

Pre-schedule consultant

If your company schedules consultants weeks or months in advance, pencil in a tentative consultant start date based on kickoff discussions and customer readiness. It's usually easier to push out a tentatively scheduled start date for the hands-on implementation work than to try and quickly schedule after a customer completes pre-reqs.

Create a high-level project plan

The SOW, data-gathering questionnaires, and kickoff meeting should provide the information needed to draft a high-level project plan:

1. Identify the tasks
   - Copy and paste the SOW tasks into a spreadsheet.
   - Add additional details, if needed. Go to the minimum detail level required for the individual task owner to understand the task and how to complete it.
   - Sequence the tasks.
   - Assign an owner to each task and deliverable.

2. Estimate how much time it will take to complete each task.

3. Email a draft of the project plan to the internal Technical team members for review and feedback.

4. Revise based on feedback.
   - If major revisions are required, schedule an internal Technical team meeting and revise in real time.

5. Email a draft of the project plan to the customer project manager for review and feedback.

6. Optimize the project plan based on the customer project manager's experience with similar projects at the company and:

- Other projects that might limit the Project team's availability.
- Holidays/vacations/training schedules.
- Balanced workload among team members.
- Whether additional time is needed for tasks dependent on others outside of the core Project team.

HIGH-LEVEL PROJECT PLAN
Sombrero Colony 2 – Cold Fusion appliance

Task Name	Owner	Start	Due	Progress	Notes
Complete pre-reqs	Bob	8/19/2224	8/21/2224	100%	
Schedule Delivery	Leya	8/24/2224	8/26/2224	50%	Awaiting approvals
Transport to customer designated location	Bob	9/06/2224	9/29/2224	0%	
Install and configure	Bob	9/30/2224	9/30/2224	0%	
Connect to power grid	Bob	10/1/2224	10/1/2224	0%	
Gradually ramp up power output	Bob	10/2/2224	10/11/2224	0%	Increase by 1 trillion kilowatt hours per day
Customer sign off	Lindsay	10/13/2224	10/13/2224	0%	

Prepare additional project documents (if needed)

Customers will sometimes request additional project docs that may not be necessary for an implementation project. These documents may or may not require significant time to create and update, so consider these requests carefully. Here are my suggestions on how to respond, based on the type of document.

Detailed project plans

The planning required to create the project plan is critical. The specific format of the plan is typically not important. Through experience, I've found that a simplified, high-level plan is sufficient for implementation projects.

Most customers will be fine with a high-level project plan in Excel or Google spreadsheet format. There are, however, a small number of project managers or sponsors who insist on a Microsoft (MS) Project plan. If a customer insists, I provide the plan in MS Project format.

The same customers who want a project plan in MS Project format typically want a detailed project plan. What these customers are typically asking for is to have installation-configuration guide steps added to the MS Project plan. I push back on this request. Managing a project plan containing tens or hundreds of installation-configuration steps that take seconds or minutes to complete can require a lot of time to update. The level of detail in the project plan should be determined by what is needed for the task owner to understand and complete the activity.

The detailed implementation project plan contains more columns than a high-level project plan. It might contain information associated with tasks' ID numbers, predecessors (dependent tasks), number of days, and graphical date information.

DETAILED PROJECT PLAN
Mars – Dome radiation protection appliance

ID	Task Name	Owner	Start	Due	Days	Progress	Pred.	August 16	23	30	September 06	13	20	27	October 04	11
	PLAN															
1.1	Schedule & prep for kickoff meeting	Fredricka	08/19/2224	08/19/2224	1	100%		■								
1.2	Complete kickoff	Team	08/27/2224	08/27/2224	1	100%			■							
	PREPARE															
2.1	Obtain power allocation approvals	Ivaan	8/30/2224	9/03/2224	4	100%				■	■					
2.2	Obtain datacenter-network approvals	Ivaan	8/30/2224	9/03/2224	4	50%				■	■					
2.3	Create test plan	Aleta	9/01/2224	9/03/2224	3	0%				■	■					
2.4	Datacenter cabling	Don	9/02/2224	9/14/2224	13	0%				■	■					
2.5	Ship appliance	Podio	9/06/2224	9/17/2224	12	0%					■	■				
	DEPLOY															
3.1	Physical install	Don	9/21/2224	9/21/2224	1	0%	2.4, 2.5						■			
3.2	Connect to Mars colony network	Don	9/22/2224	9/22/2224	1	0%							■			
3.3	Validate network-physical security	Fredricka	9/22/2224	9/22/2224	1	0%							■			
	TEST															
4.1	Existing radiation solution online	Aleta	9/27/2224	9/27/2224	1	0%								■		
4.2	Existing radiation solution offline	Aleta	9/28/2224	9/28/2224	1	0%								■		
4.3	Confirm power draw	Reyan	9/28/2224	9/28/2224	1	0%								■		
4.4	Obtain testing sign-off	Fredricka	9/29/2224	9/30/2224	1	0%								■		
	CLOSE															
5.1	Schedule KT & LL	Fredricka	9/23/2224	9/24/2224	2	0%							■			
5.2	Complete Knowledge Transfer sessions	Fredricka	9/27/2224	9/30/2224	4	0%								■		
5.3	Complete Lessons Learned meeting	Team	10/01/2224	10/01/2224	1	0%									■	
5.4	Closeout project	Fredricka	10/04/2224	10/04/2224	1	0%										■

Risk register

A risk register is a document that tracks uncertain events that, if they were to happen, could negatively or positively impact the project. (PMBOK®, Sixth Edition.)

The SOW often contains a list of assumptions and dependencies, which significantly minimizes risk. Creating and updating a risk register takes a small amount of time. I have no problem with this customer request.

RISK REGISTER
Triangulum Colony – Universal Translator Appliance

#	Description	Category	Manage	Owner	Notes
1	Space dust particles may be smaller than a nanometer	Technical	Avoid	Mike	Reconfirm force field harmonics settings prior to turnover
2	Space worms may be attracted to the nuclear core	Environment	Mitigate	Jamal, Mary	Test radiation output during post-deployment; adjust shield settings as needed
3	If project runs into delays, sunspots season will delay deployment by 2-3 months	Schedule	Accept	Adhira	
4					
5					
6					

RACI

A RACI chart shows roles for key project tasks. It's a simplified, more visual representation of the project plan. I have no problem with this customer request.

		RACI CHART				
		Mars Colony 1 – Dome radiation protection appliance				
#	Description	Mars Colony 1 Team Member	Architects	Consultant	Project Managers	Executive Sponsor
1	Obtain power allocation approvals	R	C	I	A	I
2	Obtain datacenter-network access	R	C	I	A	I
3	Complete design & test plan	C	R	C	A	I
4	Install & configure appliance	C	C	R	A	I
5	Complete tests	R	C	R	A	I
6	Document test results	R	I	R	A	I
7	Approve test results	I	I	I	A	R

R	Responsible for doing the work
A	Accountable for task success
C	Consulted for information or advice
I	Informed after task completion

Issues log

An issues log tracks problems that require assistance from someone outside of the core Project team. My preference is to not track issues separately. Management assumes the project is not going well if they see more than a few items listed. I push back on this request, as I prefer to just consider issues as action items and track them in the Chapter 9 action items log.

	ISSUES LOG Andromeda Colony 12 – Weapons Platform						
#	**Description**	**Priority**	**Logged**	**Due**	**Status**	**Owner**	**Notes**
1	Stronger than anticipated gravitational pull when the three moons are in alignment… requires 10% more power than anticipated to maintain desired orbit	Med	09/03/2224	09/09/2224	Closed	Jia	Additional solar panel added
2	Intermittent communications loss throughout the day	High	09/03/2224	09/10/2224	Closed	Cruz	Replaced transceiver array
3	Space pirates are attempting to purchase platform design/plans to identify weaknesses	High	09/03/2224	09/15/2224	In progress	Crystal	Mutant team Zeta deployed to "request" pirates immediately cease all platform-related activities
4	Internal platform temperate is 1.58 degrees warmer than anticipated	Low	09/10/2224	10/07/2224	Open	Nita	Within specifications; will investigate on next platform visit
5							
6							

Communication plan

A communication plan is a document listing communication methods and recipients. Creating and updating this plan takes a small amount of time. I have no problem with this customer request.

COMMUNICATIONS PLAN Caribbean to USA – Hurricane Destabilizer Appliances				
Description	**Frequency**	**Delivery Method**	**Distribution**	**Notes**
Kickoff	08/19/2224	Meeting	Project team and key stakeholders	Onsite
Action Items Log	Tu, Th	Email	Project team	
Risk Review	Fr	Meeting	Project team	Web conference
Status Report	Fr	Email	Project team and key stakeholders	

TOP 3 CHALLENGES

1. **Expectations: Customers often expect things that are not in the SOW. Things like custom documentation, after-hours work, and weekend work come up regularly.**

How I handle it

I have no problem providing customer-requested items that are within reason and budget, but not explicitly documented in the SOW. Some examples of common, reasonable requests include: Can you do a knowledge transfer session with our Security team? Can you help us put together an end-user communications document letting them know what's coming?

If a customer requests something time-consuming or clearly out of the SOW's scope, I discuss options and how best to respond with the Account team.

2. **Underestimating the effort and time required to obtain approvals for production environment changes in a large organization. For example, change management and security reviews can push out planned completion dates by weeks.**

   ### How I handle it

   I work with the customer's project manager to engage stakeholders during the initiate and plan phases.

3. **Accurately predicting a consultant start date for the installation/configuration activities can be a major challenge. Customer project team members are likely working on other projects. They may need a week or two to come up to speed on your product, and free up time to work on the pre-reqs.**

   ### How I handle it

   I estimate a consultant start date based on how quickly a customer completes the pre-kickoff requests. If it took a customer more than a couple of days to schedule a kickoff call or more than a week to complete pre-kickoff data gathering docs, it's unlikely pre-reqs will be completed quickly.

THINGS TO THINK ABOUT

Provide customer knowledge transfer early on.

The customer's Technical team members are typically not involved in the pre-sales process. They will likely have a different perspective and more detailed knowledge of infrastructure and applications than their team members involved in the pre-sales activities. Taking the time to demonstrate how the product works and discussing pre-reqs during the plan phase helps prevent significant issues from coming up in the middle of the project.

You're likely replacing a product or a manual process the customer has used for years.

End-users might not love the product or manual process being replaced, but they know it well. There will be some who want to leave things as they are. Make sure appropriate attention is given to their input/concerns, as their influence will grow every time there is a problem during the implementation.

Don't ignore warning signs.

If Project team members are not completing pre-reqs in a timely manner or not attending meetings, then inform the project manager and project sponsor.

Plan before starting the implementation.

Take the time to plan versus starting the work immediately. It's said every hour spent planning saves four to eleven hours of work.

Stay in your "project manager" lane.

Fight the urge to ask questions focused on helping you gain a detailed understanding of technical topics and concepts during customer meetings. Instead, focus on taking notes, keeping the team on topic, and assigning owners to action items generated during the meeting. Follow up with the consultant after the meeting with your technical questions.

Scheduling kickoff meetings.

Try to hold off committing to a kickoff call date until the customer completes and returns pre-kickoff docs. There will be times when that's not possible. For example, a key team member has a vacation scheduled to start soon, sales is pushing for a project start, customer has a deadline that requires a quick start, an executive has a quarterly bonus tied to the project, etc.

How would you handle it?

Out-of-scope requests

Omer, the project manager at a Fortune 50 company, asked me to take on the pre-req activities: scheduling meetings with his Technical team, tracking the tasks, providing weekly status reports, etc. As agreed in the SOW, the pre-reqs are the customer's responsibility.

Here's how I handled it

Customer project managers often ask the professional services project manager to own out-of-scope activities.

I've had customers ask me to help complete their internal forms, lead their pre-reqs meetings, and manage their difficult team members. These are not activities I should take on, but my mindset is the project needs to keep moving forward.

When a customer asks me to take on out-of-scope project management tasks, I ask the customer questions to understand the specific reasons and the details of what is being requested. I then weigh factors like the below before responding to the request. My goal when considering these types of requests starts with figuring out what I can say "yes" to versus saying "no."

- Is the customer request reasonable?
- What is the relationship with the customer and/or the customer's project manager?
- Is it something that can possibly come back to hurt me or my company if it doesn't work out as expected?
- Do I have the time?
- Is it required to keep the project moving?

Most of the time, I'll offer to assist, but will not take ownership of the task. However, there are some occasions when I do take ownership if it appears to be the best way to keep the project moving.

On this occasion, it only took one question for me to determine I did not need to manage the pre-reqs activities. I asked Omer why he wanted me to manage the pre-reqs tasks. Omer let me know that managing the pre-reqs looked like it would be a lot of work, and he would prefer not to be involved. I let him know my Management team would not authorize me to manage the pre-reqs activities.

He then asked, "If I manage the pre-reqs, do I still have to be involved in the implementation phase?"

"Yes. We'll need you to be involved from start to finish."

Omer then went on to ask me questions about the number of days and hours he'd have to spend on the project. Throughout the project he did as little as possible.

* Name Origin: The fictional character Homer Jay Simpson from *The Simpsons*, which surprisingly includes a good amount of science fiction-related content, fits Omer. I've never worked with a lazier and more uncaring project manager.

CHAPTER 9
EXECUTE – MONITOR – CONTROL

You have the project planned, but it's only a plan. Things won't go exactly as expected. There will be occasions when you feel like too many tasks are behind schedule or multiple team members are fighting you at every turn. This chapter will help you manage and lead the Project team through the planned and the unexpected.

KEY TERMS

Bugs: The product is not doing what's expected or not responding in the expected way.

Escalation: A request to higher-level management to help resolve an issue.

Showstoppers: Things that prevent the project from moving forward. For example, pre-reqs are not complete, security and change management approvals not obtained, firewall ports not open, consultant or key customer team member not available.

Use cases: Capturing the ways the customer plans to "use" the product. (i.e., the who, what, where, when, and how the product features will be used.)

How would you handle it?

Managing risk

In even the most well-planned projects, unexpected things can and do happen. As a project manager, it's important to be ready to handle these tricky situations when they arise.

As you read this chapter, think about how you would respond to the scenario below. At the end of the chapter, I'll share what I did and the customer's response.

The customer project manager, Rica,* has worked with you for weeks to plan a consultant's five-day onsite visit down to the hour. What happens? The assigned consultant, Evan,** was delayed for two days in Chicago due to a winter storm. There's no way he could finish five days of work in three days.

Rica asked to have Evan stay the weekend and complete the remaining tasks on Monday and Tuesday. However, he was already booked at another customer's site the following week from Monday through Friday. How would you handle the project manager's request?

KEY ACTIVITIES

70% OF YOUR TIME

For this phase, spend the majority of your time managing communication-related activities.

Manage communication

Project team status meetings
Schedule status meetings to update team members on recently completed activities, upcoming activities, and issues. Decide whether meetings should be daily, one, two, three, or four times a week based on what is needed to keep the project moving forward. Many project managers use status calls to get updates on recently completed items and issues from individual team members. My preference is to check with team members on the status of their activities between calls and provide concise updates to the team on the next call. This approach leads to more productive calls, as there is more time to discuss problems and next steps.

Meeting notes
Sending out meeting notes is a critical component of a successful project. However, project managers should not send out detailed meeting notes. If you do, few people will take the time to read them on a regular basis. Team members might even be more likely to skip meetings since they'll still receive the same amount of information as if they had attended.

I write meeting notes primarily to make my job easier, with the added bonus of keeping team members up to speed. My meeting notes provide team members with a high-level summary of topics and action items. Those same notes also provide the agenda for the next meeting, drive action items, and refresh my memory between meetings.

> **MEETING NOTES**
> **Mars Colony 1 – Dome radiation protection appliance**
>
> Hi All,
>
> Please see below for notes from yesterday's (08/27/2224) kickoff meeting
>
>   Participants: Aleta, Don, Fredricka, Ivaan, Jimmy, Podio, Reyan
>
>   Topics
>   Introductions
>   Roles and responsibilities
>   Scope
>   Key activities
>   Appliance features and capabilities
>   Change management
>   Testing
>   Pre-reqs
>   Readiness
>   Timeline/schedule
>
>   Action Items
>   Fredricka - send out kickoff meeting notes – COMPLETE
>   Ivaan - identify Security point of contact
>   Don - identify Change Management point of contact
>   Don - confirm cable connector requirements
>
> Requests
> All — Review above for individual action items
> All — Let Fredricka know if additions/corrections/updates to the above

Status reporting

Project team members are aware of status through calls and email updates. A one-page, weekly status report is an excellent way to update the project sponsor, stakeholders, and your company's Management team. Provide a weekly summary of project health (red, yellow, green), upcoming activities, and issues.

IMPLEMENTATION PROJECT MANAGEMENT

STATUS REPORT Andromeda Colony 12 – Weapons Platform			
Project Name	Andromeda Colony 12 – Weapons Platform		
Report Date	09/10/2224	**Project Health**	**GREEN**
Key week of 09/06/2224 activities			
• Test targeting systems • Test safety systems			
Anticipated week of 09/13/24 activities			
• Complete knowledge transfer sessions • Turn over final documentation • Complete lessons learned meeting			
Risk/Concerns/Roadblocks			
• None at this time			

Checking customer temperature

Regularly call and/or email the project manager and sponsor to check if they are good with how the project is going. Nothing fancy is required. Refer to the most recent status report or meeting notes, and ask if there are any concerns or issues they would like to discuss.

Manage action items

Project plans are great for planning the project, but, once it starts, the focus should shift to action items. **The importance of action items cannot be overstated.** Managing them in a proactive and consistent manner is the foundation of a successful project. Follow up with individual action item owners via calls and emails, get updates on team calls, and continually ask Project team members if they need any assistance in completing their action items.

ACTION ITEMS LOG					
Mars Colony 1 – Dome radiation protection appliance					
#	Description	Opened	Due	Status	Owner	Notes
---	---	---	---	---	---	---
1	Send out kickoff meeting notes & slide deck	08/27/2224	08/30/2224	Closed	Fredricka	
2	Identify Security point of contact	08/27/2224	08/31/2224	Open	Ivaan	Obtain estimated approval date
3	Identify Change Management point of contact	08/27/2224	09/01/2224	Open	Ivaan	Obtain estimated approval date
4	Confirm cable connector requirements	08/27/2224	08/30/2224	Open	Don	
5						
6						

Manage stakeholders

Status reports help stakeholders stay up to speed on the project. When a stakeholder makes a request or expresses a concern, make it a priority to address it in a timely manner.

Note: There is a need to rely on the customer project manager to ensure the appropriate stakeholders are identified, as you won't know their organization unless you've previously managed a project for this customer. Let the customer project manager know what job functions/departments/groups should be involved in meetings, and regularly check with the project manager to confirm the appropriate stakeholders are involved and included in weekly status report emails.

Manage escalations

Highlighting an issue or engaging an executive to assist in resolving it is sometimes necessary. While there are many things that can go wrong on an implementation project, only a couple result in an escalation on a regular basis:

Product bugs

It's impossible for a product company to duplicate every customer environment or test every customers' possible use cases. So, bugs happen. Your company's Development/Engineering/Product Management team should own bug fixes. During the implementation, your job is to make sure the fixes are prioritized and your customers receive regular progress updates.

Overdue action items

Team members are typically working on many items outside of your projects. They often don't get their action items finished by the estimated completion date. Don't make action item owners look bad on calls or in meetings by continually pointing out their missed dates. Instead, regularly and proactively follow up with individuals via emails and calls to check status and offer to assist as needed.

If an action item owner is not making progress as expected, send at least one email letting them know you'll need to let the Management team know the project may be delayed until their action item is completed.

If they continue to make insufficient progress, then do a "soft" escalation. Send an email to the action item owner requesting the planned completion date for the overdue item, copying the appropriate managers. Then, if there is still insufficient progress, escalate the issue to the customer/internal project sponsor with an email and verbal request for assistance.

20% OF YOUR TIME

Designate sufficient time to work on the below:

Update internal team members

An important part of a project manager's job is helping consultants to not only understand their specific piece of the project, but also to see the big picture. For example: whether there is another deal dependent on the success of this implementation, why certain activities that don't seem to make sense are required, political landmines, and how their work impacts others. Continually update your consultants with quick phone calls, emails, instant messages, and internal calls as needed.

Manage quarterly forecasting

Companies that charge for implementation services typically forecast how much revenue each project manager will bring in each quarter. You should provide a conservative estimate for each of your projects at the beginning of the quarter based on the limited information you have and adjust your estimate weekly.

IMPLEMENTATION PROJECT MANAGEMENT

FORECAST — Last Update: 08/27/2224

Implementation Project Name	Type	Contract Value	Remaining	Q3/2024	Q4/2024	2025	Notes
Sombrero Colony 2 – Cold Fusion appliance	Fixed	฿.000360	฿.000111	฿.000111	฿0	฿0	Project Complete
Mars Colony 1 – Dome radiation protection appliance	T&M	฿.000794	฿.000794	฿.000794	฿0	฿0	Project Complete
Mars Colony 2 – Asteroid Buster Platform	T&M	฿.000495	฿.000495	฿.000295	฿.000200	฿0	
Andromeda Colony 12 – Weapons Platform	Fixed	฿.000437	฿.000202	฿.000104	฿.000058	฿.000040	
Andromeda Colony 12 – Weapons Platform	Expenses	฿.000125	฿.000125	฿.000077	฿.000048	฿0	
Triangulum Colony 3 – Universal Translator	Fixed	฿.000437	฿.000390	฿.000100	฿.000105	฿.000185	
Caribbean to USA – Hurricane Destabilizer Appliances	T&M	฿.000140	฿.000140	฿0	฿0	฿.000140	

฿ (Bitcoin)

In 2224, a bitcoin could be worth 100 Trillion of today's dollars

Disclaimers:

- I hold a small number of bitcoins (BTC)
- This is only a fictional example.
- This is not financial advice. I am not a financial advisor.
- This example in no way suggests anyone invest/not invest in bitcoin.
- Please seek financial advice relating to your specific circumstances from a professional advisor.

Manage the schedule

Consultants will likely be scheduled on other projects before, during, and after your project completes. Work with your company's scheduler and/or project managers to confirm upcoming consultants' availability for the planned project activities. Then work with the customer project manager to identify remote and/or onsite dates for implementation activities. Keep both the customer and consultant's time zones in mind when scheduling.

Manage hours

SOW pricing is based on the estimated number of consulting hours to complete the in-scope activities. Monitor your hours and the consultant's hours to help ensure estimated hours are not exceeded. For T&M hours, you'll typically approve billable consultant hours, and regularly update customers on used and available hours. For Fixed, hours are not reported to customers. However, you'll typically still need to approve consultant hours.

Manage expenses

The project manager typically approves the consultant's travel expenses. Try to approve or reject incoming expense reports by the next business day to help avoid consultant credit card late fees. When reviewing expenses, ensure they meet your company's expense policy guidelines. Also, if the SOW references the customer's expense policy, it overrides your company's expense policy. Keep in mind when approving expense reports; a customer might question and possibly even ask for copies of a consultant's expense reports and receipts. This has happened to me a few times. If there are suspect or policy-violating charges, it may erode customer trust and result in a customer credit.

10% OF YOUR TIME

The SOW minimizes the time required for the following items, but don't neglect them altogether.

Manage risk

Identifying and keeping an eye on risk is a major activity for traditional projects. However, the SOW, prerequisites, and your company's product implementation experience greatly reduce project risks. Risks like team members leaving the company, layoffs, maternity/paternity leave, hurricanes, and snowstorms are dealt with as they come up. If there is a need to track risks, see the Chapter 8 risk register.

Manage scope change

The SOW will typically address scope and how to manage scope change requests. A single change order request can easily add days or weeks to a project. The process could include scheduling a team call to discuss the requested change, documenting the requested change, and working with the consultant to estimate the time required to complete the requested change.

If your company rejects the change order due to risk or other reasons, engage the Account team to determine the best way to inform the customer of the rejection.

If your company approves the change order and:

- No additional funding is required: Send the customer the change order form. The customer reviews the change order. Then the customer approves or rejects the change order.

- Additional funding is required: The project manager works within their company to have a quote generated. The change order and quote are sent to the customer. The customer reviews the change order and quote. The customer can then approve or reject the change.

CHANGE REQUEST			
Project Name	Mars Colony 2 – Asteroid Buster Platform		
Project Sponsor	Prime Minister Balaton	**Project Manager**	Hewstona
Requested By	General Allos	**Date**	09/13/2224
Change Description			
Testing to be expanded to include smaller asteroids between one to three meters in diameter.			
Change Reason			
Additional testing to confirm smaller targets can be acquired and destroyed			
Impact			
	Increase	Decrease	None
Schedule	TRUE	FALSE	FALSE
Budget	TRUE	FALSE	FALSE
Contracted Activities	TRUE	FALSE	FALSE
Other (Describe)			
Colonists / Systems Affected (e.g., none, one, xx%)			
None. Platform is not yet online.			
APPROVAL SIGNATURES			
Accepted (Y/N)		**Rejected (Y/N)**	
Project Sponsor		**Date**	
Project Manager		**Date**	

TOP 3 CHALLENGES

1. **Customers often want to change scope as they become more familiar with your company's product features and benefits.**

   ### How I handle it

   If it's a simple request that's low-risk and will take less than an hour or two to complete, I try to incorporate the requested change into an already planned activity, such as knowledge transfer or testing, instead of going through the formal change process. Otherwise, I follow the SOW change order process

2. **IT departments are running lean. Getting their key team members to prioritize your company's requests may be a challenge due to higher-priority activities.**

   ### How I handle it

   Most technical team members like to know in advance what is expected of them and when. So, I try to identify showstoppers during the plan phase. During the execute-monitor-control phase, I continually update the Project team members to ensure they are aware of showstoppers and their tasks/action items due date.

3. **Security requirements: No company wants to be on the news for a security breach. So, expect to provide detailed security-related information to your customers.**

How I handle it

I look to engage the appropriate Security team members during the initiate and plan phases to help ensure security requirements are identified.

THINGS TO THINK ABOUT

Time spent on various activities.

The 70%, 20%, 10% mentioned in this chapter are only guidelines based on my experience. Your actual percentages may vary based on the product being implemented and your project management style.

Project sponsor involvement might be limited.

Unless the executive sponsor considers the product as having a high strategic value, they might not be directly involved in the implementation. This is not necessarily a bad thing. Sponsors often skip calls and meetings if the project is going as expected.

Share credit.

It's a bad idea for a project manager to take credit for, or suggest they were the one responsible for, a successful project because implementation projects take a team effort. Also, it implies they're also good to accept all the blame when anything goes wrong on other/future projects, regardless of the reason. Instead, you should respond with phrases that share the credit. For example: "Thank you. It was a team effort," "Thank you. The customer was great to work with," or "Thank you. I had the easy part. Our consultant had the tough job."

Look at the spirit of the SOW.

There is a tendency for project managers to focus on doing only those tasks explicitly documented in the SOW. This can make for an unhappy customer, as there are times when an activity is implied in the SOW, but not explicitly spelled out. For example, the SOW might state global location installs, but also mention normal workday hours are 8am-5pm EST. It's natural for the customer to ask for planning discussions with key team members during their local business hours. The letter of the SOW would mean saying no to after-hours work or charging after-hours rates, but the spirit of the SOW would accommodate this request as the Sales team understood global installs.

Don't rely solely on automated meeting reminders

Technical team members can be so focused on installs, configuration or troubleshooting that they lose track of time or forget scheduled calls. Check in with key meeting participants to make sure they're prepared and still available the day before or the day of the meeting.

Don't overcommit

Project managers often try to accommodate customer requests by making same-day commitments when a customer is asking or pressing for something. If it's not urgent, provide a conservative versus aggressive completion date and time, and qualify your responses with phrases like:

- I plan to ...
- I anticipate ...
- I should be able to ...

Stuff happens

There are so many little things that can go wrong when you're managing multiple customers, projects, and consultants in various locations. No matter how well you plan, unexpected stuff will happen. Your internet connection will go down just before or in the middle of an important call you're hosting; you or consultants will have flights canceled; consultants and customers get sick. Stay calm. There's often an alternative way to handle the unexpected. If an immediate workaround is not available, apologize, work on finding a way to accomplish the planned activity in a timely manner, and figure out how to prevent it from happening again.

Protect/shield consultants

Represent your consultants when there are ad hoc customer requests like working weekends, traveling on holidays, and after hours. Discuss the request with the consultant and look for trade-offs. For example, if the request is for a consultant to work on a Sunday, get the customer to approve the following Monday as compensation time. If the request is for after hours, get the customer to approve a late start the next day.

Also, there will be times when consultants make mistakes. Do not throw them under the bus. Come up with a tactful way to handle it. Information Technology (IT) people know mistakes happen because we've all made some. If IT were easy, salaries would be a lot less. Shift the conversation from focusing on who messed up to figuring out how to fix it and prevent it from happening again.

Accommodate team members' requests

An important part of a project manager's job is to help and make it easier for Technical team members to complete their tasks and action items. If a network engineer is making good troubleshooting progress on a major issue and wants to skip a status call, then host the call without them or reschedule it. If a consultant prefers to give you verbal updates instead of updating a document, then schedule a recurring call. When a project manager continually makes a consultant's job easier, the consultant will often go the extra mile for the project manager and the project. If your team members request something reasonable, make it happen.

How would you handle it?

Managing risk

At the beginning of the chapter, you read about Rica, who had been planning a consultant's five-day onsite visit for weeks. Evan, the assigned consultant, was stuck in Chicago for two days due to a winter storm and couldn't extend his time onsite to complete the remaining tasks.

Here's how I handled it

I let Rica know the consultant was already scheduled at other customer sites for the next few weeks, but I would look to line up a consultant to work remotely with them the following week. I scheduled a consultant who was available for a couple of hours each day for the next two weeks. Rica was not happy but understood after I explained the schedule was impacted by the weather, not something I or Evan could control. It took a couple of weeks longer than originally planned to finish, but Rica still considered the project a success.

* Name Origin: The project manager reminded me of genetically engineered superhuman Khan Noonien Singh of the *Star Trek* universe played by Ricardo Montalbán. Khan was a natural leader and super smart, and so was Rica.

** Name Origin: Curtis, played by Chris Evans in the science fiction action movie *Snowpiercer,* is on a train that continually circles the planet, which has become a frozen wasteland. I felt bad for Evan. I imagined being trapped at O'Hare airport with thousands of people for two days during a snowstorm was slightly like Curtis being on the train.

CHAPTER 10
CLOSE

The final piece of the CIPM framework focuses on the steps that close out a project. These steps are sometimes seen as optional, but they're really not, so make sure you don't overlook them. They help ensure the customer is properly transitioned to your company's support or customer success organization, customer feedback is captured to help improve future projects, and the project is officially closed out.

KEY TERMS

Kudos: High praise, compliment.

Recency bias: We most easily remember and place more weight on things that have happened recently.

Rollout: Implementations are typically done in phases. The initial phase usually includes a subset of users/servers/laptops/desktops/locations/etc. Once a customer is comfortable that the product works as expected, the implementation is expanded to include additional sets of users/servers/locations/etc.

Wet signature: Physically signing a document with an ink pen.

How would you handle it?

Post-project requests

You've successfully closed out the project. Your time with the customer has ended. Or has it? If a customer runs into questions or problems after the project is closed, they'll sometimes turn to you, the project manager, for help.

As you read this chapter, think about how you would respond to the scenario below. At the end of the chapter, I'll share my take.

O'bri,* the customer project manager, calls you a few weeks after a successful implementation project has been closed out. She was great to work with and helped remove several major roadblocks.

The administrator responsible for the newly installed product unexpectedly quit yesterday. She asks if you can set up a one-hour web conference session today with the consultant who led the implementation and the person who will be taking over the duties until someone can be hired.

KEY ACTIVITIES

Inform Project team that the project is nearly complete

As the project nears completion, let the Account team know only a handful of open action items remain. Then, regularly remind team members of the remaining items in meetings, status reports, and emails.

Transition customer to support or customer success

This is a call with a customer to review how to open a case/request technical assistance, ticket response time targets, support hours, support portal/dashboard, and other relevant information.

- Schedule the meeting
- Send invite-agenda
- Facilitate the meeting

Obtain customer feedback

Satisfaction Survey

Most customer team members who take the time to reply to a feedback survey are either happy or upset with how the project went. Those in the middle don't typically respond without multiple requests to complete the survey. So, surveys are good if your primary goal is to identify projects that went well and for those that had major issues.

Keep the survey short, as in two minutes or less to complete. Don't pester survey recipients. If there is no response after a few requests, don't send a fourth request. That is, unless your manager directs you to continue following up. The survey can be sent before or after the Lessons Learned meeting.

Lessons Learned meeting

Lessons Learned meetings yield more specific and actionable information than surveys. However, they can take significantly more time and effort. The primary goal is to obtain customer feedback on what went well, areas for improvement, and next steps. This is a simple but valuable way to improve your company's implementation process.

- Schedule the meeting
- Send out invite-agenda
- Facilitate the meeting

Note: There may be times that a lessons learned meeting may not be appropriate. For example, successful projects that were simple/short, recently completed a similar project with the customer. In those cases, consider only sending a survey.

Complete your company's closeout process

The steps to officially close out a project depend on whether it is T&M or fixed price, and may vary by SOW.

T&M
Closeout may be as simple as emailing the project sponsor and project manager a summary of used hours, remaining hours, and expenses. If the in-scope work is complete and there are unused hours, send an email to the project manager and sponsor with suggestions on how to use the remaining hours. The SOW determines whether a change request approval will be needed to use the remaining hours on other tasks.

Fixed price
Closeouts typically require the customer to confirm the SOW activities are complete. Consider obtaining an electronic signature or email approval to close out the project instead of a wet signature.

Send kudos emails

Internal team members who went above and beyond deserve recognition. Send an email to them, minimally copying their manager and your manager. Summarize what they did exceptionally well and the outcome.

This email serves multiple purposes. It provides recognition and feedback, which many consultants appreciate almost as much, if not more than, a small cash reward. It also, in a subtle way, lets management know you just completed another successful project.

TOP 3 CHALLENGES

1. **Many project team members start working on or preparing for their next project before the close phase is complete. This can make wrapping up the last few items of your project difficult.**

    ### How I handle it

    I continually send reminders and follow up with Project team members to help ensure open items are being tracked and completed in a timely manner. This helps minimize the impact of team members rolling off the project.

2. **Customers raising last-minute expectations or requesting items that were not previously discussed or in the SOW.**

    ### How I handle it

    There is an acronym related to a sales strategy: "ABC." This stands for "Always Be Closing." This strategy works for implementation projects as well—always be closing the project. Throughout the project, I send regular updates and requests to team members with the intent of closing out tracked action items and identifying open items that are not being tracked.

3. Recency bias works in your favor when major problem and issues, if any, happened early in the project. It doesn't work so well if they happened near the end.

How I handle it

I ensure a Lessons Learned meeting is completed. Things that went well are the first agenda item. Discussing what went well helps to balance out and sometimes minimize what did not go well.

THINGS TO THINK ABOUT

Lessons Learned meetings are for listening.

The Lessons Learned meeting is to understand what can be improved for future projects. It's not the time for you or your company's team members to reject or explain away negative feedback. Focus on listening. Customer team members will often defend your company against their own team members when incorrect or negative comments are made.

Don't delay closing out projects.

While the project might be closed out from your company's perspective, the implementation might be part of a larger customer project. The customer might continue rolling out your company's product across their organization, test it with other products, or make customizations. Close out projects when the SOW activities are complete instead of waiting for customers to complete the rollout or extended testing. If you don't, sooner or later something will come up that results in the customer requesting additional consulting assistance at no cost. For example, a key team member leaves the company, an appliance/server dies, their environment changes, or a software upgrade causes an issue.

How would you handle it?

Post-project requests

At the beginning of this chapter, I introduced you to O'bri, the customer project manager who needed some help when their technical lead quit several weeks after the project ended. Should you schedule the implementation consultant to attend the hour-long web conference?

Here's how I handled it

This "one-hour" request could easily turn into a "one to two hours a day for weeks" request.

I let O'bri* know that I'd discuss her request with my Management team and get back to her. I sent an email to the account manager copying my manager with background and asked the account manager to follow up with O'bri. The account manager followed up with her and obtained a two-month consulting services contract to assist the customer until someone could be hired and trained to take over the administrator's duties.

* Name Origin: The project manager reminded me of Keiko O'Brien, a school teacher in the Star Trek universe played by Rosalind Chao. Like Keiko, O'bri was smart, nice and liked by everyone.

SECTION 4
CONCLUSION

By now I hope you're convinced the secret to successful product implementations is a framework light on documentation and heavy on communication. This book provides 80% of what is needed to successfully manage product implementation projects.

The other 20% will come from experience in managing projects and from gaining additional project management knowledge. I have some suggested next steps to accelerate your growth.

1. If you've not yet managed an implementation project, work with the appropriate individual in your company to be assigned to shadow another project manager or to manage a project. If you've started managing projects, manage more. Managing projects is the best, easiest way to grow your skills and help determine if you're going to like managing implementation projects.

2. Find a mentor with implementation project management experience to pair your newfound knowledge with experience. "If you don't have access to a mentor, email me (Cornell@implementationSTARS.com) when you run into challenges or have questions.

3. One of the simplest ways to take your project management skills to the next level is through my free *How to Be a STAR Implementation Project Manager* email course found at **implementationSTARS.com**.

4. Review the CIPM framework and make it work for you. For maximum effectiveness, make revisions and additions to the framework to customize it to your company's way of doing business.

INITIATE (Chapter 7)
- Review the SOW/PO/Quote
- Schedule and complete the Internal Preparation Call
- Call project sponsor
- Create a project folder

PLAN (Chapter 8)
- Schedule kickoff meeting
- Prepare for kickoff
- Send info docs in advance
- Complete kickoff meeting
- Send kickoff meeting notes
- Pre-schedule consultant
- Create a high-level project plan
- Prepare additional project documents (if needed)

EXECUTE-MONITOR-CONTROL (Chapter 9)
- Manage communication
- Manage action items
- Manage stakeholders
- Manage escalations
- Update internal team members
- Manage quarterly forecasting
- Manage the schedule
- Manage hours
- Manage expenses
- Manage risk
- Manage scope change

CLOSE (Chapter 10)
- Inform Project team that project is nearly complete
- Transition customer to support or customer success
- Obtain customer feedback
- Complete your company's closeout process
- Send kudos emails

One final thing

I hope you found the information in this book helpful. If you did, please leave a short review on Amazon. Your feedback will help other readers decide if my book would benefit them. Thanks in advance.

SECTION 5

EXTRAS

In this final section, we'll go over frequently asked questions, and seven productivity myths that might be holding back your efficiency.

FREQUENTLY ASKED QUESTIONS

Beta book readers provided excellent feedback, as well as questions. The below were asked in one form or another several times.

FAQ #1: How did you calculate 70% fewer activities are needed for PRODUCT implementation projects compared to traditional projects?

See implementationSTARS.com/seventy for details.

FAQ #2: What is the biggest challenge in managing implementation projects?

There's no universal answer. For me, it's dealing with Technical team members who continually miss their tasks/action items due date. It's a challenge for me to find the middle ground. I'm sometimes too tough or too easy on them. Some struggle with not knowing what project management skill they need to improve (self-awareness.) For most project managers, managing multiple projects simultaneously is the biggest challenge. Self-awareness and juggling projects are both covered in the *How to Be a STAR Implementation Project Manager* email course found at **implementationSTARS.com**.

FAQ #3: How do you see artificial intelligence impacting implementation project management over the next few years?

I don't see artificial intelligence having much of an impact because the implementation project manager role is so communication focused. I do, however, see the trend of companies shifting to an Everything as a Service (Xaas) mentality as having an enormous impact. The "as a Service", cloud-based subscription model will make implementations simpler, which will make a light documentation/heavy communication framework like CIPM even more applicable.

FAQ #4: Where can I get templates for the activities in Chapters 7 through 9?

There are tons of excellent free and paid templates available on the internet. You can find my favorites at implementationSTARS.com/Resources.

FAQ #5: What are your favorite movies?

The list changes from year to year, but as of publishing time, my favorite recent movies in alphabetical order are: *Alita: Battle Angel*, *Avengers: Endgame* and *Infinity Wars*, *Black Panther*, and *Mission Impossible: Fallout*.

PRODUCTIVITY MYTH BUSTERS

There is a ton of productivity and work-life balance information available in books and online. I've tried many of them. Many of the recommendations are generic enough that they're applicable to most job roles. However, there are some recommendations that seem to make sense, but are a bust when managing double-digit projects simultaneously.

Myth #1: Unplug from work when on vacation

Before going on vacation, project managers typically need to bring the backfill project manager who will handle their projects while out up to speed; inform customers of a temporary/backfill project manager; add the backfill project manager to meeting invites; add the backfill project manager to projects so they can track their time; provide backfill project manager with a document that has projects' status, contacts, project numbers, etc. This can take a lot of time. If you're going to be off for three days or fewer, it's often less work to put a message in your email signature starting a couple of weeks before your time off that you'll be off xx/xx to xx/xx and only schedule required meetings for those days.

While you're out, enable your email out-of-office autoreply indicating you'll have limited availability xx/xx to xx/xx Then spend an hour or less each day responding to "can't wait" emails, calls, and requests.

Myths #2 and #3: Don't attend calls focused on technical activities, and don't multitask

Attending technical calls allows you an opportunity to reconfirm objectives at the start of the call, ensure needed team members join the call, read attendees temperature, ensure call objectives are completed, and head off potential problems, etc.

Pretty much all other productivity advice says "don't multitask." Sitting in on technical calls is the perfect time to multitask. The key is for you to passively listen and engage as needed to keep the calls productive while working on non-critical items such as searching for customer-requested documents, moving emails from your inbox to individual email project folders, updating your timecard, etc. After you do this for a while, you'll start to listen for commitment-type words like dates and days of the week, times of the day, and phrase like, "Yes, I can do that," and, "That shouldn't be a problem." This is when you can switch back to active listening mode.

Myth #4: Call or talk in person instead of emailing

This is not practical when simultaneously managing double-digit projects. There are just too many people involved. Email is an implementation project manager's best friend. A single email can update or gather info from multiple team members, serve as a record for a decision, and remain in your inbox until the task is acted on.

Save the one-on-one calls and in-person discussions for when you're not super busy or critical, urgent, confidential, sensitive, and complicated topics.

Myth #5: Only check your email a few times a day

Project managers should continually check their email via computer and/or phone throughout the workday. Consultants working both onsite and remote often have questions or need assistance escalating issues or confirming whether a customer's ad hoc request should be worked on or not. A 15- to 30-minute delay reaching a project manager can be the difference in a consultant having a productive day versus an unproductive one. I'm often on conference calls throughout the day, so my preferred methods of communication are email and text when urgent.

Myth #6: Maintain a to-do list

When managing ten or more projects, the time spent maintaining a to-do list typically outweighs the value gained. To keep up on things that might go on a to-do list, you should develop a routine based on an email-calendar application (e.g., Outlook-Calendar, Gmail-Calendar), meeting notes, action items trackers, and your memory.

Note: You can find my personal daily routine in the *How to Be a STAR Implementation Project Manager* email course found at implementationSTARS.com.

Myth #7: Create a detailed agenda for every meeting

A detailed agenda with time limits for each agenda topic is great advice when you have one or two projects. However, it's not practical when you're simultaneously managing ten or more implementation projects and many of the meetings are recurring. Instead, look to use simple agendas like the below in your meeting invites:

- Weekly Status: updates, concerns/issues, next steps
- Project Status: what's complete, what remains, concerns/issues/roadblocks, next steps
- Internal Sync-Up: status, upcoming activities, concerns/issues/risks
- Troubleshooting: problem summary, troubleshoot, next steps

I can understand if you don't agree with some or any of the above myth busters, as you may be like me after receiving advice from Morph—skeptical. Consider trying these one at a time and see which, if any, work for you.

GLOSSARY

Action items: Small tasks assigned to an individual. Typically identified during meetings and/or associated with completion of a project plan task.

Agile: A framework that is primarily used for software development projects. Agile projects typically consist of small, regular releases of software that are based on customer or company feedback, rather than larger and less frequent releases.

Appliance/Platform: a self-contained device, sold as a product with integrated hardware and software.

Best Practices: What successful companies of similar size and industry are doing. (e.g., processes, methods, procedures).

Critical success criteria/factors: The key activities that must go well for the project to be considered a success from the customer's perspective.

Deliverables: Tangible things that will be delivered to the customer during the project. For example, project plan, test plan, working configuration, reports.

Escalation: A request to higher-level management to help resolve an issue.

Expectations: Things a customer expects but are not documented in the statement of work.

Failed project: When the project is canceled due to an issue the Implementation team could not overcome or when the customer bad-mouths a company and its products to other businesses. This usually happens due to one of the following:

- Incorrect expectations set during the sales process
- Bugs (i.e., the product is not doing what's expected or not responding in the expected way)
- Customer leadership changes, who prefer a different product
- Unsupported customer environments

Based on my experience and discussions with many other project managers, I estimate less than 2% of product implementations are considered failures.

Fixed price (Fixed): The contracted payment amount does not change regardless of the hours worked by your company to complete the project or to reach certain points within the project.

Issue: A problem that requires assistance from someone outside of the core Project team.

Kickoff: The project team's first formal meeting.

Knowledge transfer: Less formal than training, delivered by a consultant versus a trained instructor, and specific to each customers' environment.

Kudos: High praise, compliment.

Onsite: Project team members are physically working at a customer's office, data center, or a meeting location.

Prerequisites (pre-reqs): Activities to be completed or information provided prior to product installation or configuration.

Product: Commercial, pre-packaged software and/or hardware that does not require significant configuration or customization for each business customer. Typically implemented by a Professional Services or Customer Success team.

Project: Any contracted planning-deployment-implementation-optimization-knowledge transfer work assigned to the professional services/customer success group. (i.e., not considered a part of onboarding, training, or support.)

Project management: The planning and managing of activities related to completing a project.

Project managers: You, the implementation project manager, and the customer-assigned project manager work together to plan, schedule, and manage day-to-day project activities. The customer project manager typically has a fairly limited role in product implementations, as the implementation project manager has overall accountability for the project's success and activities.

Purchase Order (PO): The buyer sends your company a request to order a product or service. The document includes but is not limited to products, descriptions, quantities, prices, and payment terms. The PO and SOW contain similar information; however, the SOW is more detailed.

Quote: The quote provides customers with information needed to generate a PO (e.g., quantities, descriptions, address, contact.)

Recency bias: We most easily remember and place more weight on things that have happened recently.

Remote: Project team members can work from anywhere (e.g., your company's office, consultant's home, coffee shop)

Roadblock: Something that is stopping or slowing completion of a task.

Rollout: Implementations are typically done in phases. The initial phase usually includes a subset of users/servers/laptops/desktops/locations/etc.

Once a customer is comfortable that the product works as expected, the implementation is expanded to include additional sets of users/servers/locations/etc.

Sales: Sales team members are responsible for pre-qualifying customers, pre-sales education, identifying requirements, providing price quotes, and closing deals. The titles vary by company, but include account director, account executive, account manager, sales engineer, systems engineer, and solutions architect.

Sanitize: Removing proprietary/confidential/sensitive information and data from a document.

Scope/In scope: Specific tasks the Implementation team will perform as noted in the Statement of Work.

Showstoppers: Things that prevent the project from moving forward. For example, pre-reqs are not complete, security and change management approvals not obtained, firewall ports not open, consultant or key customer team member not available.

Simultaneous projects: A mix of active projects in various stages. (E.g., initiate, plan, execute-monitor-control, close.)

Sponsors: Internal (your company) and customer project sponsors are typically members of senior or executive management. They are the project's final decision-makers, help escalate and resolve issues, free up staff to work on projects when they are assigned elsewhere and have ultimate responsibility for the project. Also, the customer project sponsor typically funds the purchase of product and implementation services.

Stakeholders: An individual, group, or organization that may affect, be affected by, or perceive itself to be affected by a decision, activity, or outcome of a project. (PMBOK®, Sixth Edition.)

Standard/traditional project: The mythical project you typically read about in project management books. A relatively small percentage of project managers rarely, if ever, see or manage one. The scope is huge, has a seven-figure budget, 30 dedicated team members, and is the number-one priority for everyone in the company. I'm exaggerating to make a point, but the majority of project management advice assumes you want info on running large projects at a large company.

Statement of Work (SOW): Your company provides the buyer with a document that details the project activities, timeline, tasks, hours, expenses, acceptance criteria, payment schedule, etc. It's a binding contract to be signed by your company and the customer.

Subject matter experts: Individuals from your company and customers with technical expertise who provide guidance in a specific area and perform project tasks related to their area of expertise (e.g., architect, backup administrator, cloud system engineer, customer support specialist, database administrator, network engineer, security engineer.)

Note: Throughout the book we refer to subject matter experts who work at your company as consultants.

Successful project: Project success is difficult to define because it can differ from company to company. Success could be illustrated by one or more of the below:

- Project completed by the planned date, within budget, and only the planned activities were worked on
- How closely the project followed your company's project management process
- Customer management and end-users' happiness level
- High customer satisfaction survey scores
- Customer is willing to act as a reference to other potential customers

Technical leads: Typically, your company and the customer will assign technical leads to jointly provide technical expertise, develop and confirm architecture, interface with Project team members, create technical diagrams, and ensure the environment is ready for installation/configuration.

Note: Depending on product complexity, the consultant may act as both technical lead and consultant.

Throw someone under the bus: To blame another person or group for something that went wrong.

Time and Materials (T&M): Customer is billed based on the Project team members' contracted hours or days spent working on the project.

Use cases: Capturing the ways the customer plans to "use" the product. (i.e., the who, what, where, when, and how the product features will be used.)

Wet signature: Physically signing a document with an ink pen.

ACKNOWLEDGEMENTS

This book and who I am professionally are a result of literally thousands of people I've worked with over the last 30 years. I would like to thank a handful of mentors who were especially instrumental in helping me along the way:

- Alan Randolph: strong work ethic
- Lew Wagner: business writing and seeing the big picture
- Mathew Palakal: office politics
- Sameer Bhatia: continual stretch assignments

Made in the USA
Monee, IL
30 December 2020